Other subjects available online

CW00555137

# WHAT'S IN THE BOOK?

## Thousands of maths questions

## Everything in the book is online at www.lbq.org

www.lbq.org

$A = \begin{pmatrix} 2 & -1 \\ -3 & -2 \end{pmatrix}$

## Use the book to help

find, plan, prepare, rehearse, personalise AND... learn how you can use lbq.org in your classroom for...

## Whole Class Teaching
Teach mode turns any question into a slide - ideal for whiteboard or touchscreen display

## Ad hoc Questioning
Instantly create and send questions to your class

## Self Paced Tasks
Set Questions Sets for your class to work through at their own pace.

# Teach

If you've got a whiteboard or touch screen you can turn any question into a whole class teaching resource

Wherever you see the 'Teach' symbol you're one click away from turning a question or Question Set into a teaching resource.

TEACH ICON

Each question can be an ideal teaching point

Annotate to explain, model and work as an example

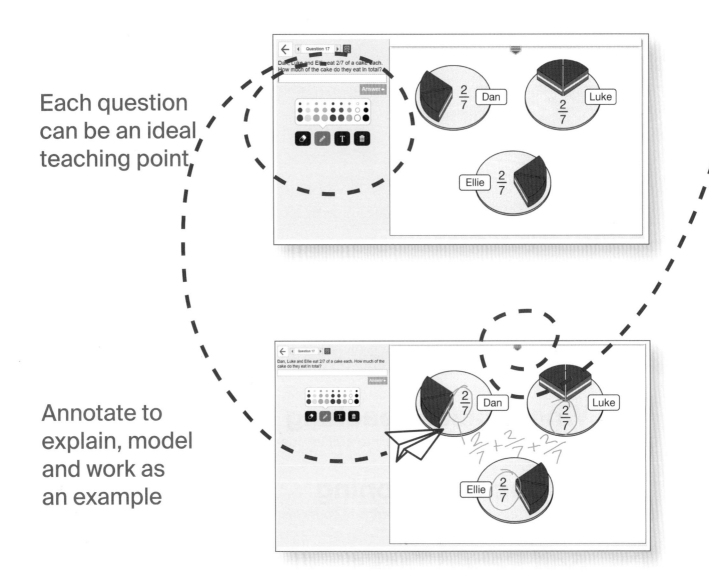

Use the pull down pad to construct your own questions

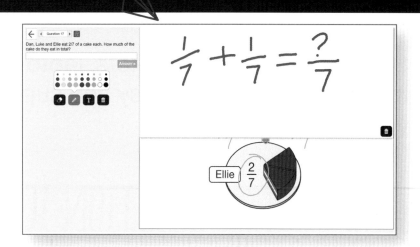

Work through multiple questions like a slide show

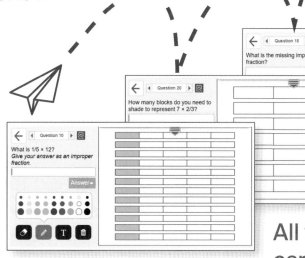

All the questions are grouped into carefully scaffolded sets and provide great support for progression from simple practice to mastery

TEACH MODE IS **FREE** TO ALL REGISTERED USERS

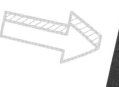

**Register FREE at lbq.org**

USE THE BOOK TO HELP FIND, ORGANISE AND REHEARSE QUESTIONS TO INCLUDE IN YOUR LESSONS

# Ad hoc Questioning*

Have your pupils got tablets, laptops, chromebooks, PCs...? lbq.org makes it fast, easy and super-productive to engage your whole class.

## THE AD HOC QUESTION ICON

(always available top right on lbq.org) is your gateway to asking questions:  at anytime...
of everyone...
about anything...

Are your class struggling with a challenging question?

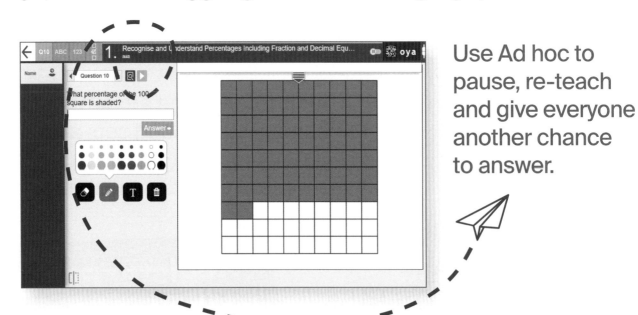

Use Ad hoc to pause, re-teach and give everyone another chance to answer.

*AD HOC QUESTIONING IS **FREE** FOR 60 DAYS!

**Register FREE at lbq.org**

# Want to build on an existing lbq.org question?

Annotate to explain, model, modify and extend...

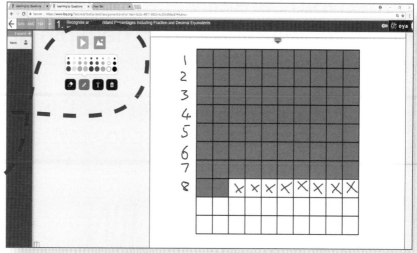

 ...and use Ad hoc to send as a new question to your class.

## Or just make your own questions on the fly?

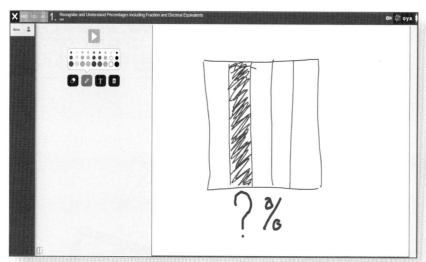

Use our teach tools to create the right question at the right time.

Write a question, draw a question or just ask a question...

**... and forget 'hands up'.** With Ad hoc questioning everyone answers, every question, every time!

USE THE BOOK TO CHOOSE, ORGANISE AND REHEARSE QUESTIONS TO TEACH IN YOUR CLASSES

# Self Paced Tasks*

Lbq.org is built on tens of 1000's of questions - questions grouped into carefully scaffolded sets to provide structured support for learning and to help pin point problems.

When you click the 'Start' button you're opening the door for your class to start answering questions

START BUTTON

Pupils connect with a simple code and start receiving questions straight away.

Enter your code

6 l d

using Lbq Tasks

App Store | Google play | Get it from Microsoft

*Subscription required for self-paced and ad hoc tasks after initial trial. Teach mode remains free.

**Register FREE at lbq.org**

# And when your pupils hit a challenging question...

... you'll know about it.

So drill down to see every answer - be right on top of every misconception.

And try the question again with ad hoc questioning.

Then pause, intervene, explore, explain, model with the 'Teach' features.

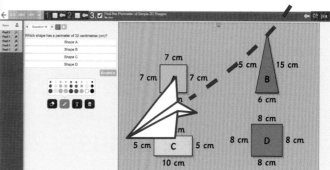

Everything in this book is online at www.lbq.org

USE THE BOOK TO PICK THE RIGHT TASKS FOR YOUR CLASS AND BE READY FOR INTERVENTION OPPORTUNITIES.

# Other titles in the series

## Learning by Questions

| PRIMARY KS2 | SECONDARY KS3 |
|---|---|
| **Maths**<br>Year 3 Mathematics Primary Question Sets<br>Year 4 Mathematics Primary Question Sets<br>Year 5 Mathematics Primary Question Sets<br>Year 6 Mathematics Primary Question Sets | **Maths**<br>Year 7 Mathematics Primary Question Sets*<br>Year 8 Mathematics Primary Question Sets*<br>Year 9 Mathematics Primary Question Sets* |
| **Science**<br>Years 3&4 Science Lower KS2 Question Sets*<br>Years 5&6 Science Upper KS2 Question Sets* | **Biology**<br>KS3 Biology Question Sets* |
|  | **Chemistry**<br>KS3 Chemistry Question Sets* |
| **English**<br>Years 3&4 English Lower KS2 Question Sets*<br>Years 5&6 English Upper KS2 Question Sets* | **Physics**<br>KS3 Physics Question Sets* |
|  | **English**<br>KS3 English Question Sets* |
| **US Math**<br>Grades 4&5 Mathematics Question Sets** | **US Math**<br>Grades 6&7 Mathematics Question Sets** |

\* Available January 2019
\*\* Available summer 2019

See www.lbq.org/books for title availability

# Understanding a question set

Question set title

Topic & sub topic

Curriculum objective

Quick Search Reference number

Default feedback to students

Block number & title

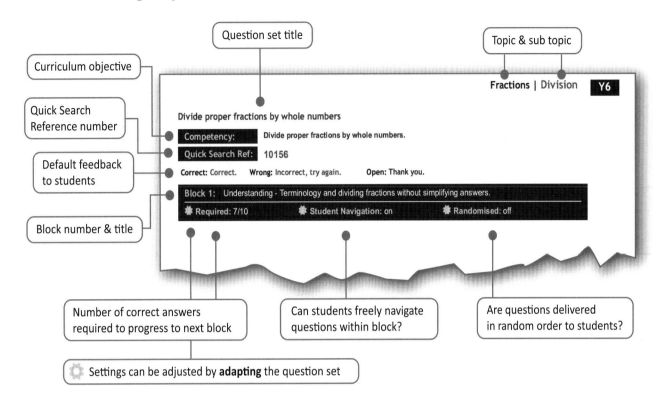

Fractions | Division    Y6

Divide proper fractions by whole numbers

**Competency:**    Divide proper fractions by whole numbers.

**Quick Search Ref:**    10156

**Correct:** Correct.    **Wrong:** Incorrect, try again.    **Open:** Thank you.

**Block 1:**    Understanding - Terminology and dividing fractions without simplifying answers.
✹ Required: 7/10    ✹ Student Navigation: on    ✹ Randomised: off

Number of correct answers required to progress to next block

Can students freely navigate questions within block?

Are questions delivered in random order to students?

⚙ Settings can be adjusted by **adapting** the question set

# Understanding a question

Question

Answer options, green for correct.

Question Type

Multiple Choice

Numeric

2/5 (answers required)

Text

Sort in order

True or False

Yes or No

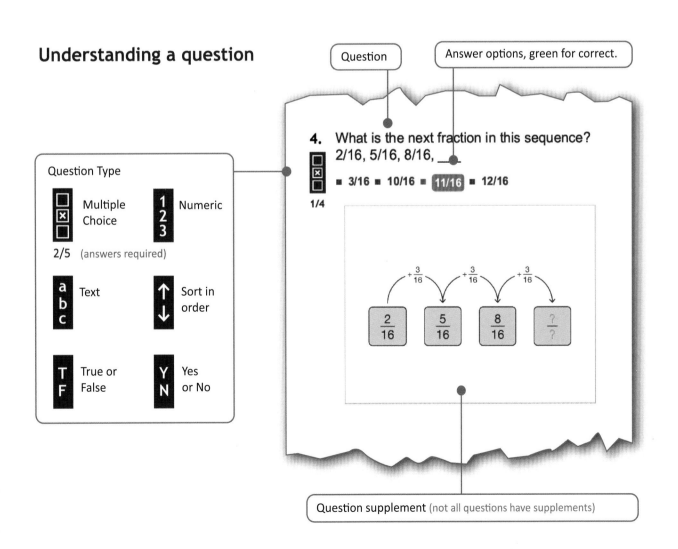

**4.** What is the next fraction in this sequence?
2/16, 5/16, 8/16, ___

■ 3/16  ■ 10/16  ■ 11/16  ■ 12/16

1/4

$+\frac{3}{16}$  $+\frac{3}{16}$  $+\frac{3}{16}$

$\frac{2}{16}$    $\frac{5}{16}$    $\frac{8}{16}$    $\frac{?}{?}$

Question supplement (not all questions have supplements)

# Finding a Question set from this book on the LbQ Platform

The **year group, topic** and **sub topic** classifications used in this book relate directly to those used on the LbQ platform.
The fastest way to find a specific question set on lbq.org is via the **Quick Search Reference Number** (e.g. 10654).
To further refine a search select a distinctive **keyword** from the question set title or competency.

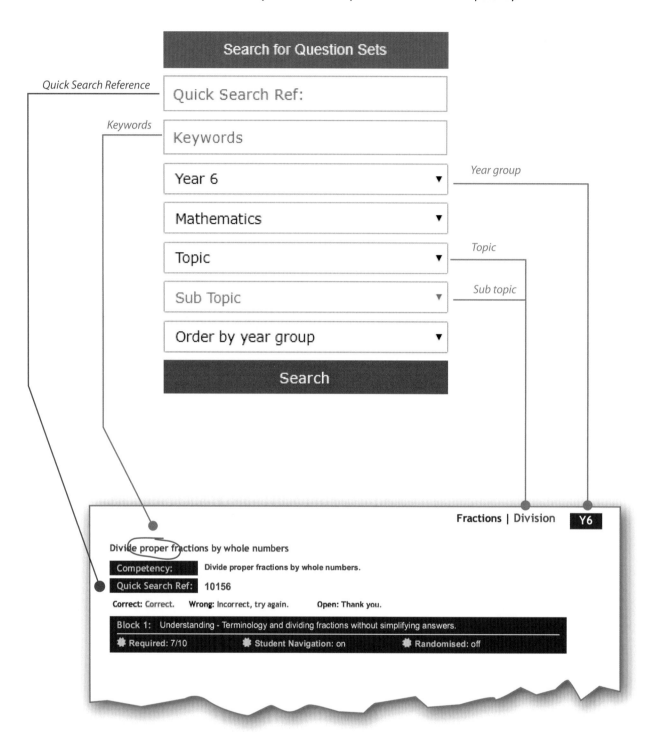

**Note:** The question sets detailed in this book are correct at time of compilation (July 2018) and correspond directly to the question sets as published on www.lbq.org.

Owing to the nature of www.lbq.org we will from time to time extend, update or modify the question sets published there, which will give rise to discrepancies between this book and the online resources.

# Topic Directory | Y7

# Place Value

## Compare and Order

## Rounding and Estimation

# Compare and order integers

**Competency:** Order positive and negative integers, use the number line as a model for ordering of the real numbers; use the symbols =, ≠, <, >, ≤, ≥.

**Quick Search Ref:** 10146

Correct: Correct.     Wrong: Incorrect, try again.     Open: Thank you.

**Level 1:** Understanding - Comparing integers using the number line.

✱ **Required:** 7/10       ✱ **Student Navigation:** on       ✱ **Randomised:** off

**1.** An integer is:
- a number less than zero. ■ a whole number.
- a number between zero and one. ■ a positive number.

1/4

**2.** Which two integers are negative numbers?
- -0.47 ■ 6 ■ -17 ■ 0 ■ -¼ ■ -289

2/6

**3.** Is the following statement true or false?
1 is less than -9.
- True ■ False

1/2

**4.** Which integer has a greater value, 14 or -6?
- -6. ■ 14

1/2

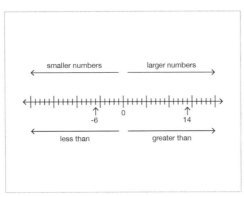

**5.** Which integer has a greater value, -13 or -18?
- -13 ■ -18

1/2

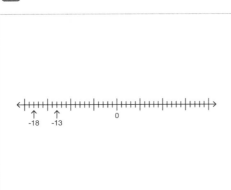

**6.** Arrange the following integers in descending order (largest first):
-9, 12, -33.
- 12 ■ -9 ■ -33

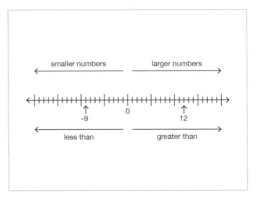

**7.** Arrange the following integers in ascending order (smallest first):
-12, 99, -100.
- -100 ■ -12 ■ 99

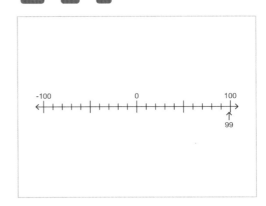

**8.** Which number has a greater value, -1 or 1?
- -1 ■ 1

1/2

**Level 1:** *cont.*

9. Is the following statement true or false?
0 is greater than -8?

■ True   ■ False

1/2

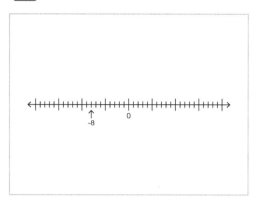

10. Arrange the following integers in ascending order (smallest first):
-14, -17, 3.

■ -17   ■ -14   ■ 3

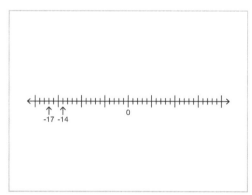

**Level 2:** Fluency - Using symbols to compare and order integers.

✱ **Required:** 7/10   ✱ **Student Navigation:** on
✱ **Randomised:** off

11. Arrange the symbols in the following order:
greater than;
less than;
equal to;
not equal to;
less than or equal to;
greater than or equal to.

■ >   ■ <   ■ =   ■ ≠   ■ ≤   ■ ≥

12. Select the symbol that makes the statement true:
-501 ___ -502

■ >   ■ <

1/2

$$-501 \quad \boxed{\phantom{>}} \quad -502$$

13. Select the expression that shows:
*n* is greater than or equal to 4.

■ n > 4   ■ n = 4   ■ n < 4   ■ n ≥ 4   ■ n ≤ 4   ■ n ≠ 4

1/6

14. Arrange the cities in descending order according to their average temperature (highest first).

■ San Francisco   ■ London   ■ Beijing   ■ Moscow
■ Ottawa

| average daytime temperatures in January | |
| --- | --- |
| city | temperature (°C) |
| Ottawa | -10 |
| Beijing | -3 |
| London | 4 |
| Moscow | -7 |
| San Francisco | 11 |

15. Which symbol makes the following statement true?
14 ___ 16

1/4   ■ =   ■ >   ■ ≠   ■ ≥

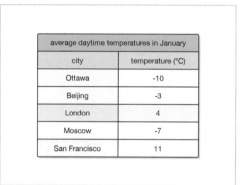

$$14 \quad \boxed{\phantom{=}} \quad 16$$

**Level 2:** *cont.*

**16.** Arrange the cities in descending according to their height above sea level (highest first).

- Vienna ▪ Edinburgh ▪ London ▪ Amsterdam
- Azerbaijan ▪ Jericho

| city | height above/below sea level (m) |
|------|------|
| Jericho | -258 |
| Azerbaijan | -28 |
| London | 14 |
| Edinburgh | 47 |
| Amsterdam | -2 |
| Vienna | 170 |

**17.** Which **two** integers are greater than -6 and less than or equal to -3?

▪ -8 ▪ 5 ▪ -6 ▪ -3 ▪ 0 ▪ -5

2/6

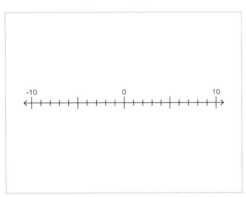

**18.** If integer *d* is less than -5, which **three** statements are correct?

▪ d ≥ -5 ▪ d ≤ -4 ▪ d < 5 ▪ d ≤ -6

3/4

**19.** Which symbol makes the following statement true?

-12,354 ___ -12,345

▪ <

**20.** Arrange the companies in descending order according to their annual profit (greatest first).

- Edith's Electricals ▪ Gulab's Garage ▪ Tanya's Taxis
- Ahmed's Antiques ▪ Harry's Hotel

| companies in Sumtown annual profits | |
|------|------|
| company | profit/loss (£) |
| Harry's Hotel | -31,583 |
| Tanya's Taxis | 8,579 |
| Ahmed's Antiques | -9,480 |
| Gulab's Garage | 27,084 |
| Edith's Electricals | 46,912 |

**21.** Paul is thinking of an integer less than 1. Suzie is thinking of an integer greater than or equal to -1 but not equal to 0. Paul and Suzie are thinking of the same number. What is the number?

▪ -1

**22.** What integer is halfway between -7 and 5 on the number line?

▪ -1

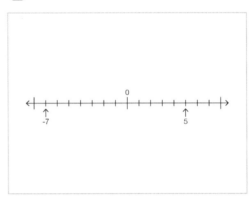

**23.** How many integers are greater than -10 and less than 10?

▪ 19

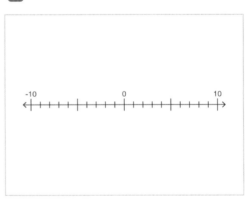

**24.** Ralph says, "The closer a number is to zero, the smaller it is". Is Ralph correct? Explain how you know.

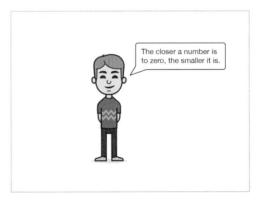

The closer a number is to zero, the smaller it is.

**Level 3:** *cont.*

**25.** Jordan is thinking of an integer. His number is greater than -2 but less than or equal to 4. How many different possible integers could Jordan be thinking of?

- 6

**26.** Carla is thinking of an odd, one-digit number that is less than -3. If her number is not next to -8 on the number line, what is Carla's number?

- -5

**Level 4:** Problem Solving - Involving positive and negative integers.

❋ **Required:** 6/6    ❋ **Student Navigation:** on
❋ **Randomised:** off

**27.** The numbers are arranged in descending order. What is the value of *n*?

- 4

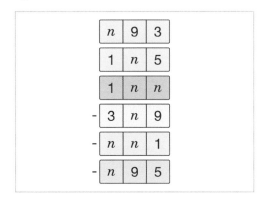

**28.** Using the digit cards 5 to 9, what is the closest number you can make to -70,000?

- -69875  - -69,875

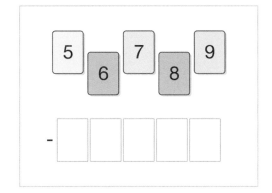

**29.** The number in each square is exactly halfway between the two numbers in the circles on either side of it. What number goes in the highlighted square?

- -1

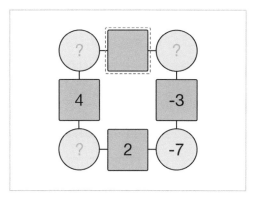

**30.** What is the largest even negative number you can make using the digit cards?

- -13498  - -13,498

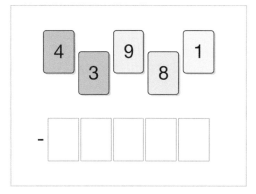

**31.** C is a number which is twice as far from B as it is from A. There are two possible values for C. What is the sum of the two possible values?

- -26

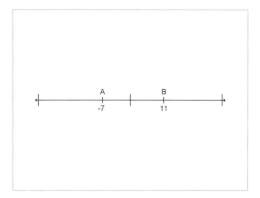

**32.** Use the digit cards to complete the grid.

 All the rows, columns and diagonals must add up to the same number.

What three-digit number do you end up with on the top row?

■ 834

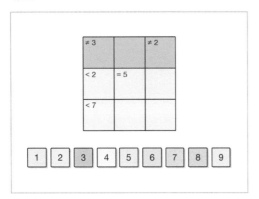

# Round to any number of decimal places

**Competency:** Round numbers and measures to an appropriate degree of accuracy.

**Quick Search Ref:** 10138

Correct: Correct.    Wrong: Incorrect, try again.    Open: Thank you.

**Level 1:** Understanding - Rounding to any number of decimal places.

✱ **Required:** 7/10    ✱ **Student Navigation:** on    ✱ **Randomised:** off

1.  In the number 93.4718, what digit is in the second decimal place?

   ▪ 7

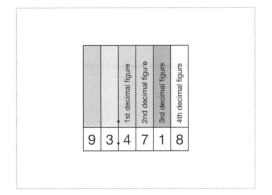

2.  When rounding 0.325 to two decimal places (2 d.p.), would you round up or down?

   ▪ up   ▪ down

   1/2

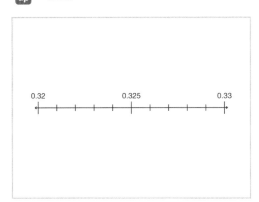

3.  What does the ≈ symbol mean?

   ▪ equal to   ▪ approximately equal to   ▪ less than
   ▪ not equal to

   1/4

4.  What is 4.2374 rounded to two decimal places (2 d.p.)?

   ▪ 4.24

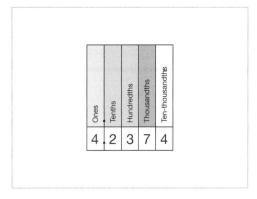

5.  What is 0.04681 rounded to three decimal places (3 d.p.)?

   ▪ 0.047

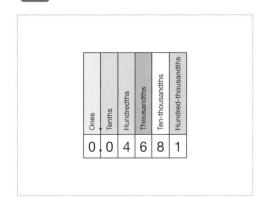

6.  Round 2.645741 to four decimal places (4 d.p.).

   ▪ 2.6457

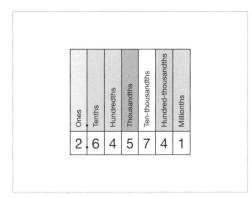

**Level 1:** *cont.*

**7.** Round 0.007341 to three decimal places (3 d.p.).

1
2
3   ■ 0.007

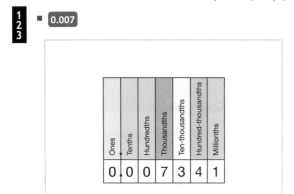

**8.** In the number 86.5081, what digit is in the third decimal place?

1
2
3   ■ 8

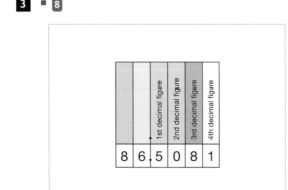

**9.** What is 0.625 rounded to two decimal places (2 d.p.)?

1
2
3   ■ 0.63

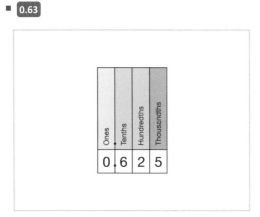

**10.** Round 36.1872 to three decimal places (3 d.p.).

1
2
3   ■ 36.187

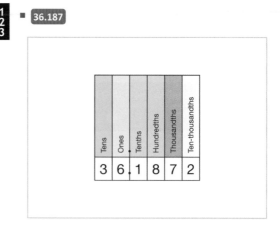

**Level 2:** Fluency - Rounding in context including misconceptions carrying digits and trailing zeros.

✲ **Required:** 7/10    ✲ **Student Navigation:** on
✲ **Randomised:** off

**11.** A bureau de change lists £1.00 as currently being worth $1.28592. What is the exchange rate rounded to two decimal places?

1
2
3   ■ 1.29

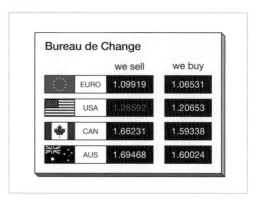

**12.** Round 57.6038 to two decimal places.

a
b
c   ■ 57.60

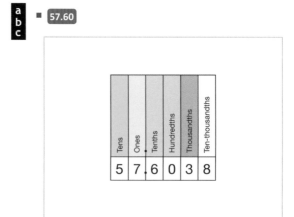

**Level 2:** *cont.*

**13.** The population of China in 2017 can be written as 1,388.2 million rounded to one decimal place. Give the population of the United States in millions rounded to one decimal place.

■ 326.5

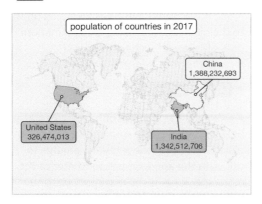

**14.** A chocolate bar contains 30.014 grams of fat per 100 grams of chocolate. How should the manufacturer show this figure rounded to one decimal place on the nutritional label?

■ 30.0

| nutrition information | | | |
|---|---|---|---|
| | per 100 g | per bar (35g) | *reference intakes |
| energy | 2150 kj 515 kcal | 752 kj 180 kcal | 8400 kj 2000 kcal |
| total fat<br>   saturated fat | | 10.5 g | 20 g |
| carbohydrate<br>   of which sugars | 60.5 g 48.0 g | 21.0 g 16.5 g | 260 g 90 g |
| fibre | 1.2 g | 0.4 g | - |
| protein | 7.2 g | 2.5 g | 50 g |
| salt | 0.5 g | 0.2 g | 6 g |

**15.** Use short division to find the value of 1 ÷ 7 rounded to three decimal places.

■ 0.143

**16.** Pi (π) is a number with an infinite amount of decimal places. Rounded to eleven decimal places pi is 3.14159265359. What is pi rounded to four decimal places?

■ 3.1416

**17.** Manchester United's attendance can be written as 1.43 million rounded to two decimal places. How would you write Arsenal's attendance in millions rounded to two decimal places?

■ 1.14

| Premier League 2016/2017 top 3 total match attendances | |
|---|---|
| Manchester United | 1,430,502 |
| Arsenal | 1,139,177 |
| West Ham United | 1,082,464 |

**18.** John is going to Spain for his holidays and will get €1.09919 for every £1.00 he exchanges. How many Euros will he get for each pound rounded to two decimal places?

■ 1.10

**19.** Round 83.4963 to two decimal places.

■ 83.50

**20.** Use short division to find the value of 14.388 ÷ 3 rounded to two decimal places.

■ 4.80

$$3\overline{)14.388}$$

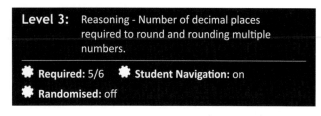

**Level 3:** Reasoning - Number of decimal places required to round and rounding multiple numbers.

✸ **Required:** 5/6    ✸ **Student Navigation:** on
✸ **Randomised:** off

21. Salma says, "If you round a number to more decimal places, you **always** get a more accurate answer". Is Salma correct? Explain your answer.

    [a b c]

22. Finn and Caitlin each calculate 1.7483 + 4.6319 to 1 decimal place in two different ways:
    Finn rounds the numbers to one decimal place and then adds them; Caitlin adds the numbers first and then rounds her answer to one decimal place. Which method gives the most accurate answer?

    1/3

    - rounding both numbers then adding
    - adding the numbers then rounding
    - they are both the same

23. The following numbers have been rounded to a given number of decimal places. Which statement is correct?

    1/4

    - 0.2869 ≈ 0.28 (2 d.p.)  ▪ 16.3891 ≈ 16.39 (3 d.p.)
    - 23.5028 ≈ 23.5 (2 d.p.)  ▪ **36.996 ≈ 37.00 (2 d.p.)**

24. Isabel is dividing an 8 metre wall into 7 equal sections, so she calculates 8/7.
    How many decimal places will she need to calculate to get an answer correct to the nearest centimetre?

    ▪ **3**

25. Round each number to the given number of decimal places and then arrange the values in **descending** order.

    ▪ 3.80052 (3 d.p.)  ▪ 3.76124 (1 d.p.)  ▪ 3.78717 (2 d.p.)
    ▪ 3.78791 (3 d.p.)  ▪ 3.78742 (4 d.p.)

26. Lillie thinks of a number with two decimal places. When she rounds her number to one decimal place it doubles in size. What is Lillie's number?

    ▪ **0.05**

**Level 4:** Problem Solving - Errors in approximation.

✸ **Required:** 5/5    ✸ **Student Navigation:** on  -
✸ **Randomised:** off

27. Danielle's height is 151.3 centimetres (cm) rounded to one decimal place. What is Danielle's smallest possible height in centimetres?
    *Don't include the units in your answer.*

    ▪ **151.25**

28. Jasper and Cody each think of a different number with two decimal places. When rounded to one decimal place, they both get 1.6. What is the largest possible difference between their original numbers?

    ▪ **0.09**

29. A bag of crisps has a mass of 25 grams (g) rounded to the nearest gram. What is the minimum possible mass of six bags of crisps?
    *Include the units g (grams) in your answer.*

    ▪ **147 g**  ▪ **147 grams**

30. A rectangle has a length of 6.2 metres (m) and a height of 4.5 m, both rounded to one decimal place. Jamie uses these measurements to calculate the approximate perimeter of the rectangle. What is the maximum possible difference, in metres, between Jamie's estimate and the actual perimeter of the rectangle?
    *Include the units m (metres) in your answer.*

    ▪ **0.2 m**  ▪ **0.2 metres**

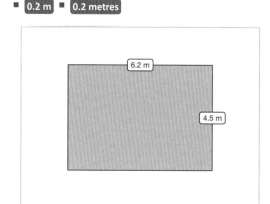

31. A wall has a width of 3 metres (m) and a height of 4 m, both rounded to the nearest integer. Danny estimates the area of the wall as 12 m². What is the maximum amount that Danny may have overestimated the size of the wall by in metres squared?
    *Don't include the units in your answer.*

    ▪ **3.25**

# Round to any number of significant figures

**Competency:** Round numbers and measures to an appropriate degree of accuracy. For example, to a number of significant figures.

**Quick Search Ref:** 10172

**Correct:** Correct. **Wrong:** Incorrect, try again. **Open:** Thank you.

**Level 1:** Understanding - Rounding numbers and decimals to any number of significant figures.

⚙ **Required:** 7/10 ⚙ **Student Navigation:** on ⚙ **Randomised:** off

---

**1.** In the number 5,129, what is the most significant figure?

■ **5** ■ **1** ■ **2** ■ **9**

1/4

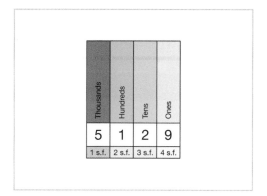

**2.** What does the ≈ symbol mean?

■ **equal to** ■ **greater than** ■ **not equal to**
■ **approximately equal to**

1/4

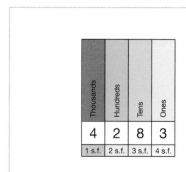

**3.** Round 4,283 to one significant figure (1 s.f.).

■ **4,000** ■ **4000**

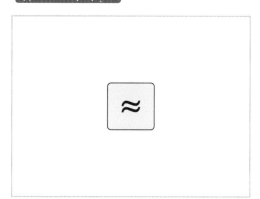

**4.** Round 76,925 to one significant figure (1 s.f.).

■ **80000** ■ **80,000**

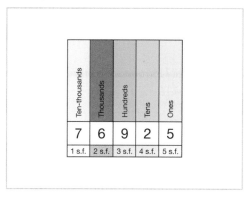

**5.** A concert has an audience of 23,487. What is the attendance rounded to two significant figures (2 s.f.)?

■ **23000** ■ **23,000**

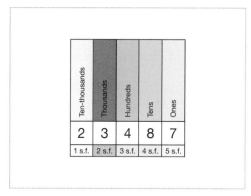

**6.** In 2017 the population of Fiji was approximately 902,547. What is this rounded to three significant figures (3 s.f.)?

■ **903,000** ■ **903000**

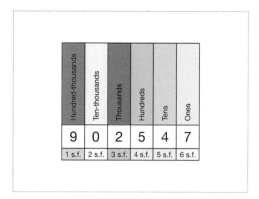

---

**7.** What is 37.82915 rounded to four significant figures (4 s.f.)?

**1**
**2**
**3**

▪ 37.83

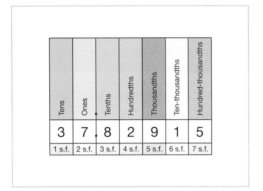

**8.** What is 5.23716 rounded to three significant figures (3 s.f.)?

**1**
**2**
**3**

▪ 5.24

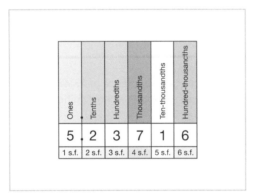

**9.** What is 8.4517 rounded to three significant figures (3 s.f.)?

**1**
**2**
**3**

▪ 8.45

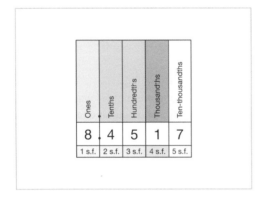

**10.** Round 427,813 to two significant figures (2 s.f.).

**a**
**b**
**c**

▪ 430,000  ▪ 430000

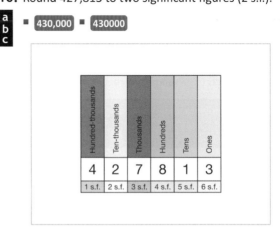

**Level 2:**  Fluency - Dealing with misconceptions involving zeros.

✱ **Required:** 7/10  ✱ **Student Navigation:** on
✱ **Randomised:** off

**11.** In the number 0.07462, what is the most significant figure?

**1**
**2**
**3**

▪ 7

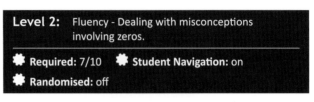

**12.** Round 0.0871 to two significant figures (2 s.f.).

**1**
**2**
**3**

▪ 0.087

**13.** What is 83.017925 rounded to four significant figures (4 s.f.)?

 ▪ 83.02

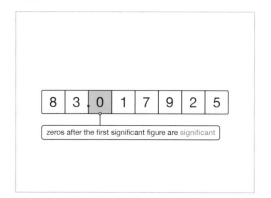

**14.** Round 12.50694 to four significant figures (4 s.f.).

 ▪ 12.51

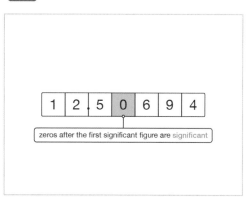

**15.** Round 0.0507826 to three significant figures (3 s.f.).

 ▪ 0.0508

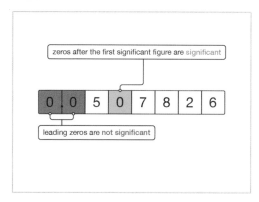

**16.** What is 63.90418 rounded to four significant figures (4 s.f)?

 ▪ 63.90

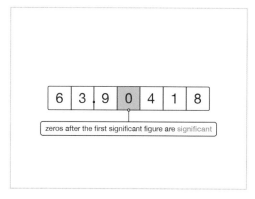

**17.** Round 2.79681 to three significant figures (3 s.f.).

  ▪ 2.80

**18.** Round 53.9723 to three significant figures (3 s.f.).

  ▪ 54.0

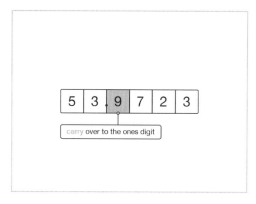

**19.** What is 0.0085126 rounded to three significant
 figures (3 s.f.)?

■ 0.00851

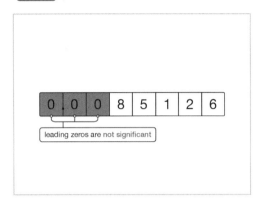

**20.** What is 0.4026719 rounded to four significant
 figures?

■ 0.4027

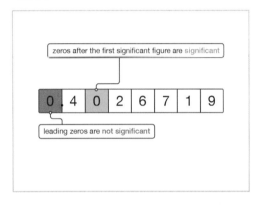

**Level 3:** Reasoning - Rounding multiple numbers and
approximation.

🌼 **Required:** 5/7    🌼 **Student Navigation:** on
🌼 **Randomised:** off

**21.** The following numbers have been rounded to a
given number of significant figures. Which
statement is correct?

1/5

■ 26,742 ≈ 26,000 (2 s.f.)  ■ 5.23971 ≈ 5.24 (4 s.f.)
■ 0.0681 ≈ 0.07 (2 s.f.)  ■ 23.6981 ≈ 23.70 (4 s.f.)
■ 0.043076 ≈ 0.04308 (3 s.f.)

**22.** Round each number to the given number of
significant figures and then arrange the values
in **descending** order (largest first).

■ 326.2 (2 s.f.)  ■ 327.961 (3 s.f.)  ■ 327.974 (5 s.f.)
■ 327.94 (4 s.f.)

**23.** The attendance at a music festival was 135,000
when rounded to three significant figures. What is
the **largest** possible attendance?

■ 135499

**24.** Tamsin says, "If you round a number to more
significant figures, you **always** get a more accurate
answer". Is Tamsin correct? Explain your answer.

**25.** Charlie and Salma each think of a different five-
digit number. When Charlie rounds his number to
two significant figures it is the same as Salma's
number rounded to to three significant figures.
What is the maximum possible difference between
Charlie's and Salma's numbers?

■ 549

**26.** A race course measures 10 kilometres (km) when
rounded to two significant figures (2 s.f.). What is
the minimum possible length of the race course in
metres (m)?
*Don't include the units in your answer.*

■ 9,950  ■ 9950

**27.** The length of a square field is 200 metres rounded

to two significant figures (2 s.f.). Leighton uses this
measurement to calculate the approximate
perimeter of the field as **800 metres**. What is the
maximum possible difference between Leighton's
calculation and the actual perimeter?
*Give your answer in metres (m) but don't include
the units in your answer.*

▪ **20**

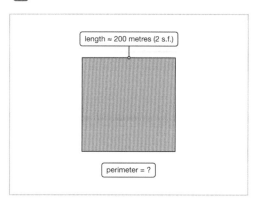

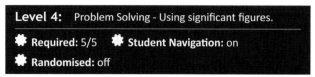

**31.** What is the 99th most significant figure in the
answer to the calculation, 16 ÷ 11?

▪ **5**

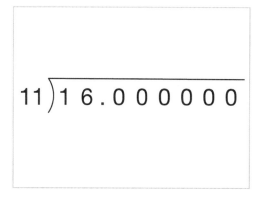

**32.** Keith and Cath each think of a two digit integer.
The sum of their numbers is 110. When rounded
to one significant figure, Keith's number is three
times larger than Cath's. What number is Cath
thinking of?

▪ **25**

**Level 4:** Problem Solving - Using significant figures.

✿ **Required:** 5/5    ✿ **Student Navigation:** on
✿ **Randomised:** off

**28.** The weight of a crate is rounded to four significant
figures (4 s.f.). Sarah uses this value to estimate
the weight of 3 crates as 37.8 kilograms. What is
the **minimum possible weight** of **one crate** in
grams?

▪ **12,595** ▪ **12595**

**29.** Each fencing panel measures 2.4 metres (m) wide,
to two significant figures (2 s.f.). What is
the maximum number of panels you would need
to make a fence measuring **at least** 100 metres
long?

▪ **43**

**30.** Using each digit card **once**, make an even number
that is 4,000 rounded to one significant figure (1
s.f.) and 3,800 rounded to two significant figures (2
s.f.).

▪ **3784** ▪ **3,784**

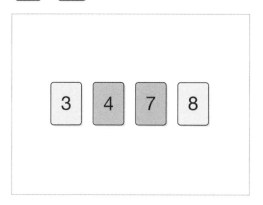

# Addition and Subtraction

Decimals

Written Methods Subtraction

Inverse Operations

Negative Numbers

# Add and subtract integers and decimals of any size

**Competency:**   Add and subtract integers and decimals using formal written methods.

**Quick Search Ref:**   10222

**Correct:** Correct.   **Wrong:** Incorrect, try again.   **Open:** Thank you.

**Level 1:**   Understanding - Add and subtract large integers and decimals up to 2 d.p.

✱ **Required:** 7/10      ✱ **Student Navigation:** on      ✱ **Randomised:** off

---

**1.**  Work out the answer to 4,152 + 783.

 ▪ **4,935** ▪ **4935**

```
    4 1 5 2
+     7 8 3
```

**2.**  What is 2,187 - 1,469?

 ▪ **718**

```
    2 1 8 7
-   1 4 6 9
```

**3.**  Add 1.2 and 0.7.

 ▪ **1.9**

```
      1 . 2
+     0 . 7
```

**4.**  Complete the calculation 3.74 - 1.25 = ____.

 ▪ **2.49**

```
    3 . 7 4
-   1 . 2 5
```

**5.**  What is the sum of 198.72 and 63.5?

 ▪ **262.22**

```
    1 9 8 . 7 2
+     6 3 . 5 0 ─┐
                 place -
                 holding zero
```

**6.**  What is the sum of 2.63 and 1.84?

▪ **4.47**

```
      2 . 6 3
+     1 . 8 4
```

**Level 1: cont.**

**7.** Find the difference between 23.9 and 15.87.

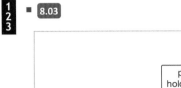 ▪ 8.03

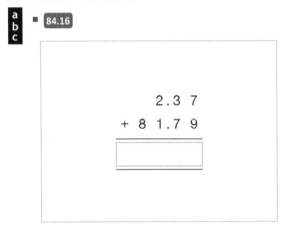

place - holding zero

2 3.9 0

− 1 5.8 7

**8.** Add 2.37 and 81.79.

abc ▪ 84.16

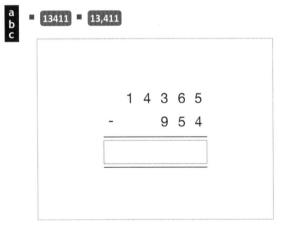

2.3 7

+ 8 1.7 9

**9.** Work out the answer to 14,365 - 954.

abc ▪ 13411 ▪ 13,411

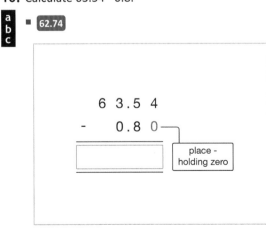

1 4 3 6 5

− 9 5 4

**10.** Calculate 63.54 - 0.8.

abc ▪ 62.74

6 3.5 4

− 0.8 0

place - holding zero

**Level 2:** Fluency - Add and subtract decimals up to 3 decimal places including questions in context.

✹ **Required:** 8/10 ✹ **Student Navigation:** on
✹ **Randomised:** off

**11.** Find the sum of 4.527, 65.3 and 95.015.

abc ▪ 164.842

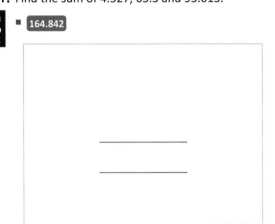

**12.** Work out the difference between 10.87 and 27.

abc ▪ 16.13

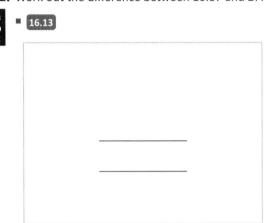

**13.** Calculate 0.53 - 0.007.

abc ▪ 0.523

**Level 2:** *cont.*

**14.** What is 147.8 + 9.508?

a
b
c
■ 157.308

**15.** What is the missing number?

a
b
c

1.304 + _____ = 3.895

■ 2.591

**16.** Harriet wants to buy a book for £4.37. She has £1.48. How much more does she need?
*Include the £ sign in your answer.*

a
b
c

■ £2.89

**17.** Joanna and Michelle take part in a race. Joanna completes the race in 25.3 seconds and Michelle finishes 1.97 seconds later. How long does it take Michelle to finish the race?
*Include the units (s) seconds in your answer.*

a
b
c

■ 27.27 s   ■ 27.27 seconds

**18.** A new exercise book costs 65p, a ruler costs 39p and a calculator costs £3.98. How much do the three items cost in total?
*Include the £ sign in your answer.*

a
b
c

■ £5.02

**19.** Danny has three apples which weigh 74.2 grams, 0.089 kilograms and 91.06 grams. What is the total weight of the three apples?
*Include the units g (grams) in your answer.*

a
b
c

■ 254.26 grams   ■ 254.26 g

**Level 2: cont.**

**20.** What is the missing number?

25.3 + 521.5 + ___ = 684.12.

▪ 137.32

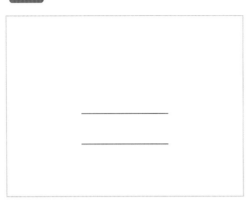

**Level 3:** Reasoning - Add and subtract integers and decimals of any size.

✿ **Required:** 5/5   ✿ **Student Navigation:** on
✿ **Randomised:** off

**21.** What is the next number in the sequence?

1
2
3
▪ 2.5

| 12.1 | 8.9 | 5.7 | |

**22.** Zoe has added the two decimals together as shown. Explain the mistake that she has made.

a
b
c

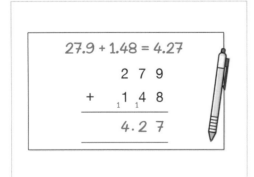

27.9 + 1.48 = 4.27

```
    2 7 9
+   1 4 8
  1   1
  4 . 2 7
```

**23.** Find the difference between A and B.

1
2
3
▪ 9.425

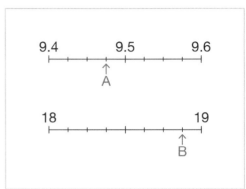

**24.** Amy uses the inverse calculation 142.07 + 86.89 to calculate the missing value in the following equation:

a
b
c

142.07 - ___ = 86.89.
Is Amy correct? Explain your answer.

**25.** Nancy is using a calculator to subtract a number from 6.42. She accidentally presses the addition key and gets the answer 8.2.
What should her answer have been?

1
2
3

▪ 4.64

**Level 4:** Problem Solving - Add and subtract integers and decimals of any size.

✿ **Required:** 5/5   ✿ **Student Navigation:** on
✿ **Randomised:** off

**26.** What is the value of *a*?

1
2
3

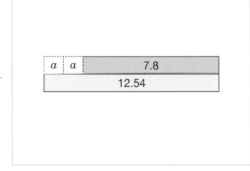

| *a* | *a* | 7.8 |
| 12.54 | | |

Level 4: *cont*.

**27.** Add the values in two circles to find the answer to the box between them. What number is missing from the blank box?

▪ 7.88

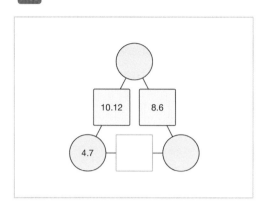

**28.** Calculate the 3 missing digits. What is the closest number to 400 you can make with the 3 digits?

 ▪ 384

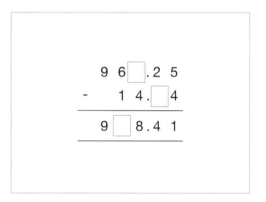

**29.** Add the numbers in two side-by-side boxes to find the answer to the box above. What number goes in the striped box?

▪ 0.24

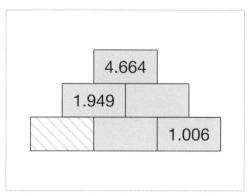

**30.** In the calculation shown *a*, *b* and *c* each represent a single digit. What is the answer to the calculation?

▪ 7.43

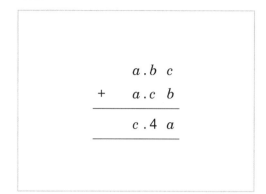

$$
\begin{array}{r}
a\,.\,b\ \ c \\
+\quad a\,.\,c\ \ b \\
\hline
c\,.\,4\ \ a \\
\hline
\end{array}
$$

# Use Inverse Operations Involving the Addition and Subtraction of Decimals

**Competency:** Recognise and use relationships between operations including inverse operations.

**Quick Search Ref:** 10319

Correct: Correct.   Wrong: Incorrect, try again.   Open: Thank you.

**Level 1:** Understanding - Inverse operations up to 2 decimal places.

**Required: 7/10**    **Student Navigation: on**    **Randomised: off**

**1.** What is the inverse operation of subtraction?

☐
☒   ▪ addition  ▪ multiplication  ▪ subtraction  ▪ division
☐

1/4

**2.** What number makes the following statement correct?

_____ + 7.3 = 14.5

▪ 7.2

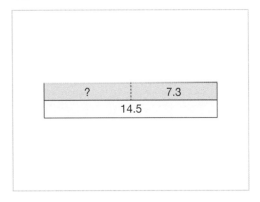

**3.** What is the missing number?

26.2 = _____ + 15.6

▪ 10.6

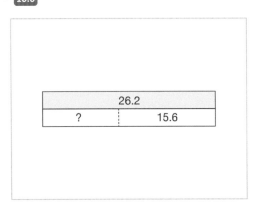

**4.** What number completes the following statement?

_____ - 6.7 = 3.2

▪ 9.9

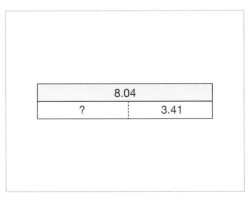

**5.** What number would satisfy the following number statement?

_____ + 3.41 = 8.04

▪ 4.63

**6.** What is the missing number?

36.37 = _____ - 13.65

▪ 50.02

Level 1: *cont.*

**7.** What would satisfy the following number statement?
123
_____ - 27.49 = 84.51

▪ 112

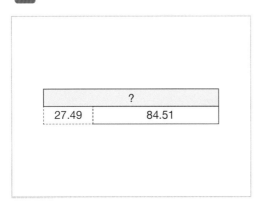

**8.** What number makes the following statement correct?
abc
63.04 = _____ + 47.32

▪ 15.72

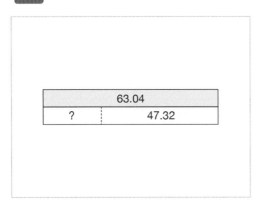

**9.** What is the missing number?
abc
_____ - 38.4 = 91.3

▪ 129.7

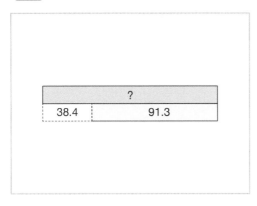

**10.** What number would satisfy the following number statement?
123
27.25 + _____ = 60

▪ 32.75

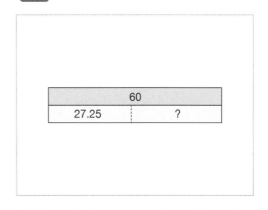

**Level 2:** Fluency - Inverse operations with decimals and in context.

✸ Required: 7/10  ✸ Student Navigation: on
✸ Randomised: off

**11.** What is the missing amount?
abc
£2.95 + £5.76 = £3.00 + _____.
*Include the pound sign (£) in your answer.*

▪ £5.71

**12.** Saffy is 0.07 metres (m) too short for a ride at the funfair. If the height restriction is 1.24 m, how tall is she?
abc
*Include the units m (metres) in your answer.*

▪ 1.17 m  ▪ 1.17 metres

**13.** What number makes the statement correct?
123
23.8 + _____ = 91.2

▪ 67.4

**14.** What is the missing number?
123
_____ - 47.6 = 75.8.

▪ 123.4

**15.** After spending £63.28 on shopping, Jenny has £182.19 left in her bank account. How much did she have in her bank before she went shopping?
abc

▪ £245.47

**16.** What is the missing amount?
abc
£67.24 - £39.89 = _____ - £40.00.
*Include the pound sign (£) in your answer.*

▪ £67.35

**17.** What number makes the following statement correct?
abc
643.15 = _____ + 427.3

▪ 215.85

**Level 2: cont.**

**18.** What is the missing mass in kilograms (kg)?

24.13 kg + _____ = 23.24 kg + 17.01 kg.
*Don't include the units in your answer.*

■ 16.12

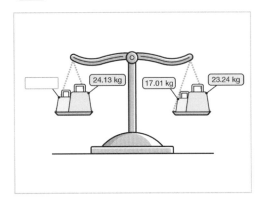

**19.** What number would satisfy the following number statement?

a
b
c

321.45 + _____ = 500

■ 178.55

**20.** What is the missing number?

a
b
c

83.19 = _____ - 49.32

■ 132.51

**Level 3:** Reasoning - Using inverse operations to compare values and work out missing numbers.

✻ **Required:** 5/5   ✻ **Student Navigation:** on
✻ **Randomised:** off

**21.** Work out the missing numbers and then sort the equations by value in descending order (largest first).

↑
↓

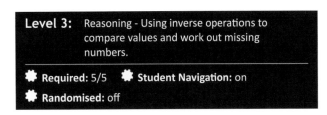

■ _____ + 37.3 = 58.8   ■ 37.3 = 58.6 - _____
■ 64.9 = 43.8 + _____   ■ 14.6 = _____ - 6.4

**22.** Lisa writes the following in her notepad:

a
b
c

**63.2 + 24.9 = 24.9 + 63.2**
**so 63.2 - 24.9 = 24.9 - 63.2**
Is Lisa correct? Explain your answer.

**23.** Three of the number sentences have the **same missing number**. Which is the odd one out?

☐
☒
☐

1/4

■ A) _____ - 17.5 = 26.3   ■ B) 80.7 = 36.9 + _____
■ C) _____ + 25.4 = 69.2   ■ D) 11.8 = _____ - 31.9

**24.** Select two numbers to complete the number sentence:

☐
☒
☐

2/6

_____ + _____ + 17.8 = 50

■ 9.7   ■ 16.1   ■ 17.8   ■ 22.5   ■ 32.2   ■ 40.3

**25.** 274.41 + _____ = 641.38

a
b
c

Freya says the missing number is 915.79.
Adil says the missing number is 366.97.
Joe says they are both wrong.
Who is correct? Explain your answer.

**Level 4:** Problem Solving - Using the inverse operation to solve puzzles and multi-step word problems.

✻ **Required:** 5/5   ✻ **Student Navigation:** on
✻ **Randomised:** off

**26.** Calculate the total of *a*, *b* and *c*.

1
2
3

$c + b = a$
$a + 35.5 = 123.2$
$45 - b = 16.7$

■ 175.4

**27.** In the number pyramid, each number is the sum of the two numbers directly below it. What number goes in the highlighted box?

1
2
3

■ 16.7

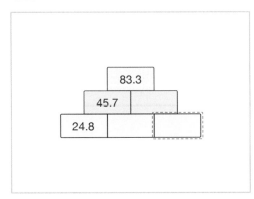

**28.** If the number in each square is the sum of the two numbers in circles on either side of it, what number goes in the blank square?

1
2
3

■ 506.1

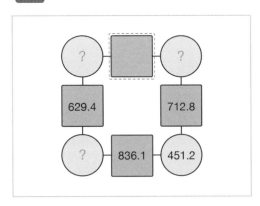

**29.** Find the value of *z*:

1
2
3

$x + 68.3 = 100$
$x + y = 50$
$x - y + z = 25$

■ 11.6

**30.** Gemma wants to raise £100 for charity.

She raises £24.75 from a sponsored run and **£10.41 more** from a sponsored swim. Her sponsored cycle earns £24.40 less than the total from the first two events. How much more does she need to reach her target?
*Include the pound sign (£) in your answer.*

▪ £4.58

# Use Inverse Operations Involving the Addition and Subtraction of Integers and Decimals

**Competency:** Recognise and use relationships between operations including inverse operations.

**Quick Search Ref:** 10219

Correct: Correct.    Wrong: Incorrect, try again.    Open: Thank you.

**Level 1:** Understanding - Inverse operations up to 3-digit numbers.

✸ **Required:** 7/10    ✸ **Student Navigation:** on    ✸ **Randomised:** off

1. What is the inverse operation of addition?

   ▪ Addition ▪ Multiplication ▪ Subtraction ▪ Division

   1/4

2. What number makes the following statement correct?
   _____ + 23 = 75.

   ▪ 52

3. What is the missing number?
   92 = _____ + 36.

   ▪ 56

4. What number completes the following statement?
   _____ - 67 = 32.

   ▪ 99

5. What number would satisfy the following number statement?
   _____ + 278 = 731.

   ▪ 453

6. What is the missing number?
   327 = _____ - 184.

   ▪ 511

7. What number would satisfy the following number statement?
   _____ - 236 = 489.

   ▪ 725

8. What number makes the following statement correct?
   645 = _____ + 278.

   ▪ 367

9. What number would satisfy the following statement?
   293 + _____ = 628.

   ▪ 335

10. What is the missing number?
    _____ - 384 = 529.

    ▪ 913

**Level 2:** Fluency - Inverse operations up to 4 digit numbers, with decimals and in context.

✸ **Required:** 7/10    ✸ **Student Navigation:** on
✸ **Randomised:** off

11. What number makes the following statement correct?
    _____ + 2,836 = 7,324.

    ▪ 4488 ▪ 4,488

12. If the scales shown are balanced, what is the missing mass in grams (g)?
    *Include the units in your answer.*

    ▪ 75 g ▪ 75 grams

13. What is the missing number?
    _____ - 3,185 = 5,376.

    ▪ 8561 ▪ 8,561

14. What is the missing amount?
    £2.95 + £5.76 = £3.00 + _____.
    *Include the pound sign (£) in your answer.*

    ▪ £5.71

15. What number makes the statement correct?
    23.8 + _____ = 91.2

    ▪ 67.4

**Level 2:** *cont.*

**16.** The perimeter of the field is 420 metres (m).
a The length of the field is 137 metres.
b What is the **width** of the field?
c *Include the units m (metres) in your answer.*

   ▪ 73 metres ▪ 73 m

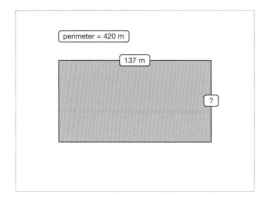

**17.** If the scales shown are balanced, what is the
a missing mass in grams (g)?
b *Include the units in your answer.*
c

   ▪ 248 grams ▪ 248 g

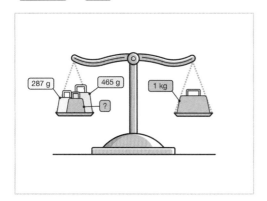

**18.** What is the missing number?
1 _____ - 47.6 = 75.8.
2
3 ▪ 123.4

**19.** After spending £63.28 on shopping, Jenny has
a £182.19 left in her bank account. How much did
b she have in her bank before she went shopping?
c *Include the units £ (pounds) in your answer.*

   ▪ £245.47

**20.** What is the missing amount?
a £67.24 - £39.89 = _____ - £40.00.
b *Include the pound sign (£) in your answer.*
c

   ▪ £67.35

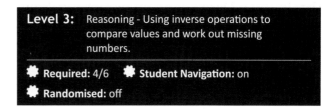

**Level 3:**  Reasoning - Using inverse operations to
          compare values and work out missing
          numbers.

✿ **Required:** 4/6    ✿ **Student Navigation:** on
✿ **Randomised:** off

**21.** Work out the missing numbers and then sort the
↑ equations by value in descending order (largest
↓ first).

   ▪ [_____ + 37.3 = 58.8]  ▪ [37.3 = 58.6 - _____]
   ▪ [64.9 = 43.8 + _____]  ▪ [14.6 = _____ - 6.4]

**22.** Lisa writes the following in her notepad:
a **63.2 + 24.9 = 24.9 + 63.2**
b **so 63.2 - 24.9 = 24.9 - 63.2**
c Is Lisa correct? Explain your answer.

**23.** I think of a number, add 245, subtract 484, and
1 finally add 376. My answer is 600. What is my
2 number?
3
   ▪ 463

**24.** Darren used a calculator to add a number to 1,234
a but he accidentally pressed the **subtract button**
b instead of add. If Darren got the answer 777, what
c answer would he have got if he had pressed the
**add button**?

   ▪ 1,691 ▪ 1691

**25.** Joshua says, "The sum of two numbers is always
a greater than the difference".
b Is Joshua correct? Explain your answer.
c

**26.** Three of the number sentences have the **same
□ missing number**. Which is the odd one out?
☒
□ ▪ A) _____ - 17.5 = 26.3 ▪ B) 80.7 = 36.9 + _____
1/4 ▪ C) _____ + 25.4 = 69.2 ▪ D) 11.8 = _____ - 31.9

**Level 4:** Problem Solving - Using the inverse operation to solve puzzles and multi-step word problems.

✸ **Required:** 5/5   ✸ **Student Navigation:** on
✸ **Randomised:** off

**27.** Use the digits 1 to 5 to make the following statement correct. What number goes in the green box?
1
2
3
__ __ __ - __ __ = 316

▪ 341

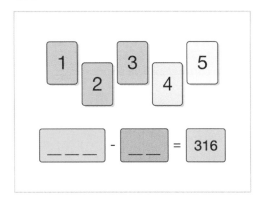

**28.** In the number pyramid, each number is the sum of the two numbers directly below it. What number goes in the highlighted box?
1
2
3

▪ 16.7

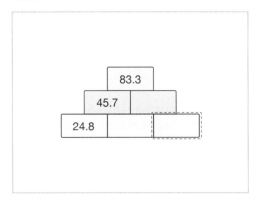

**29.** If the number in each square is the sum of the two numbers in circles on either side of it, what number goes in the blank square?
1
2
3

▪ 506.1

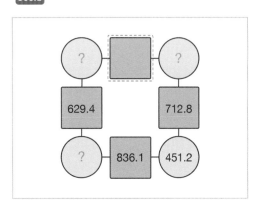

**30.** Find the value of $z$:
1
2
3
$x + 68.3 = 100$
$x + y = 50$
$x - y + z = 25$

▪ 11.6

**31.** Jason took 25% of his T-shirts to the charity shop, but then he went to a concert and bought another 2 two. If Jason now owns 14 T-shirts, how many did he have originally?
1
2
3

▪ 16

# Use inverse operations involving the addition and subtraction of integers

**Competency:**    Recognise and use relationships between operations including inverse operations.

**Quick Search Ref:**    10317

Correct: Correct.    Wrong: Incorrect, try again.    Open: Thank you.

**Level 1:**    Understanding - Inverse operations up to 3-digit numbers.

✹ Required: 7/10    ✹ Student Navigation: on    ✹ Randomised: off

1.  What is the inverse operation of addition?

    ▪ addition  ▪ multiplication  ▪ subtraction  ▪ division

    1/4

2.  What number makes the following statement correct?
    _____ + 23 = 75

    ▪ 52

3.  What is the missing number?
    92 = _____ + 36

    ▪ 56

4.  What number completes the following statement?
    _____ - 67 = 32

    ▪ 99

5.  What number satisfies the following number statement?
    _____ + 278 = 731

    ▪ 453

6.  What is the missing number?
    327 = _____ - 184

    ▪ 511

7.  What number satisfies the following number statement?
    _____ - 236 = 489

    ▪ 725

8.  What number makes the following statement correct?
    645 = _____ + 278

    ▪ 367

9.  What number would satisfy the following statement?
    293 + _____ = 628

    ▪ 335

10. What is the missing number?
    _____ - 384 = 529

    ▪ 913

**Level 2:**    Fluency - Inverse operations up to 4-digit numbers and in context.

✹ Required: 7/10  ✹ Student Navigation: on
✹ Randomised: off

11. What number makes the following statement correct?
    _____ + 2,836 = 7,324

    ▪ 4488  ▪ 4,488

12. Adam travelled 1,375 kilometres (km). He stopped after 463 km and then stopped again after a further 235 km. How much of the journey did he have left at his second stop?
    *Include the units km (kilometres) in your answer.*

    ▪ 677 kilometres  ▪ 677 km

13. If the scales shown are balanced, what is the missing mass in grams (g)?
    *Include the units g (grams) in your answer.*

    ▪ 75 g  ▪ 75 grams

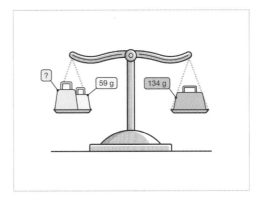

14. What is the missing number?
    _____ - 3,185 = 5,376

    ▪ 8561  ▪ 8,561

**Level 2:** *cont.*

**15.** If there is 268 millilitres (ml) of juice left in a bottle
**a**  after 379 ml has been used, how much was in the
**b**  bottle originally?
**c**  *Include the units ml (millilitres) in your answer.*

   ■ 647 millilitres  ■ 647 ml

**16.** What is the length of the base of the isosceles
   triangle?

   ■ 44 cm  ■ 58 cm  ■ 130 cm  ■ 304 cm

1/4

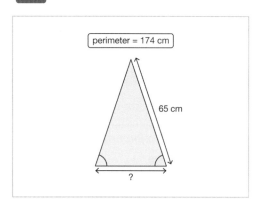

**17.** The perimeter of the field is 420 metres (m) and
**a**  its length is 137 metres. What is the **width** of the
**b**  field in metres?
**c**  *Include the units m (metres) in your answer.*

   ■ 73 m  ■ 73 metres

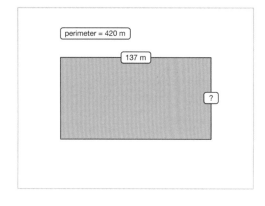

**18.** If the scales shown are balanced, what is the
**a**  missing weight?
**b**
**c**  *Include the units g (grams) in your answer.*

   ■ 248 g  ■ 248 grams

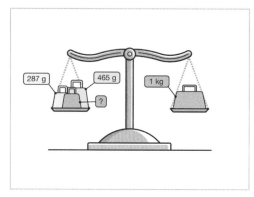

**19.** What is the missing number?
**a**  _____ + 7,865 = 9,000
**b**
**c**  ■ 1,135  ■ 1135

**20.** What is the missing number?
**a**  785 + _____ + 2,312 = 5,028 - 411
**b**
**c**  ■ 1,520  ■ 1520

**Level 3:** Reasoning - Using inverse operations to
compare values and work out missing
numbers.

✸ **Required:** 5/5   ✸ **Student Navigation:** on
✸ **Randomised:** off

**21.** Fran thinks of a number. She adds 245, subtracts
**1**  484, and finally adds 376. Her answer is 600.
**2**  What is Fran's original number?
**3**
   ■ 463

**22.** Darren used a calculator to add a number to 1,234
**a**  but he accidentally pressed the **subtract button**
**b**  instead of add. If Darren got the answer 777, what
**c**  answer would he have got if he had pressed the
   **add button**?

   ■ 1691  ■ 1,691

**23.** Joshua says, "The sum of two numbers is always
**a**  greater than the difference".
**b**  Is Joshua correct? Explain your answer.
**c**

**24.** Work out the missing numbers and then arrange
↑  the equations according to the value of the
↓  missing number, in ascending order (smallest first).

   ■ 1,111 + _____ = 4,224   ■ 2,509 = _____ - 620
   ■ 4,761 = 1,428 + _____   ■ 2,323 = 5,720 - _____

**25.** Select **two** numbers to complete the number
   sentence:
   6,358 + _____ - _____ = 9,927

2/5  ■ 4,292  ■ 723  ■ 3,569  ■ 2,845  ■ 7,082

**Level 4:** Problem Solving - Using the inverse operation to solve puzzles and multi-step word problems.

✱ **Required:** 5/5      ✱ **Student Navigation:** on
✱ **Randomised:** off

26. Use the digits 1 to 5 to make the following
    statement correct. What number goes in
    the highlighted box?
    __ __ __ - __ __ = 316

    ▪ 341

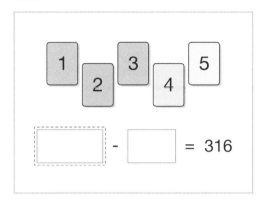

27. Jayne has 2 metres (m) of red ribbon and 3 metres
    each of green and blue. She uses 1,075 millimetres
    (mm) of red, 625 mm of green and 1,265 mm of
    blue ribbon to wrap 5 presents. How many
    millimetres of ribbon does she have left?
    *Include the units mm (millimetres) in your answer.*

    ▪ 5,035 mm   ▪ 5,035 millimetres   ▪ 5035 millimetres
    ▪ 5035 mm

28. In an addition pyramid, the number in each box is
    the sum of the numbers in the two boxes directly
    below it. What number goes in the striped box?

    ▪ 889

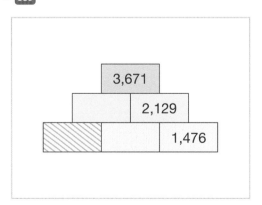

29. Jason took 25% of his t-shirts to the charity shop,
    but then he went to a concert and bought another
    2. If Jason now owns 14 t-shirts, how many did he
    have originally?

    ▪ 16

30. Find the total of the 4 missing digits from the
    calculation.

    ▪ 18

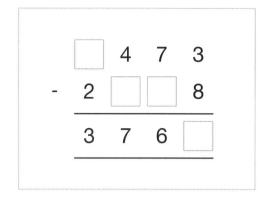

# Add and subtract negative numbers

**Competency:** Use the four operations, including formal written methods, applied to integers, decimals, proper and improper fractions and mixed numbers, all both positive and negative.

**Quick Search Ref:** 10263

Correct: Correct.    Wrong: Incorrect, try again.    Open: Thank you.

**Level 1:** Understanding - Meaning of negative and positive, sorting numbers and understanding calculations.

✱ **Required:** 7/10    ✱ **Student Navigation:** on    ✱ **Randomised:** off

1. What is a **negative** number?

■ A number less than zero.  ■ A number greater than zero.
■ A number equal to zero.
1/3

2. Which statement best describes zero?
■ Zero is a positive number.  ■ Zero has no value.
■ Zero is a negative number.
1/3

3. Helen eats 9 grapes out of a packet.
What number represents this?

■ -9

4. Sort the numbers in **ascending** order (smallest first).
■ -84 ■ -38 ■ -9 ■ -1 ■ 22 ■ 51 ■ 72

5. Select the **two** statements that are true.

■ Two like signs together (+/+ or -/-) mean you should add.
■ Two unlike signs together (+/- or -/+) mean you should subtract.
2/4
■ Two like signs together (+/+ or -/-) mean you should subtract.
■ Two unlike signs together (+/- or -/+) mean you should add.

6. Select the correct calculation for the following expression:
3 + -5 =
1/4  ■ 3 + 5  ■ 3 - 5  ■ -3 + 5  ■ -3 - 5

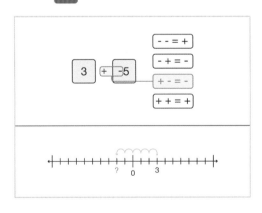

7. What is the correct calculation for the following expression:
5 - -7 =
1/4  ■ 5 - 7  ■ -5 - 7  ■ 5 + 7  ■ -5 + 7

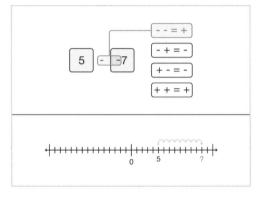

8. Sort the numbers in **descending order** (largest first).
■ 7 ■ 4 ■ 1 ■ 0 ■ -3 ■ -5 ■ -9

9. Zarah throws away 7 carrots.
What number represents this?
a b c  ■ -7

10. What is the correct calculation for the following expression:
-6 + 7.
1/4  ■ -6 - 7  ■ 6 + 7  ■ 6 - 7  ■ -6 + 7

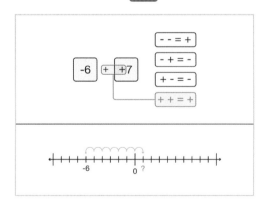

**Level 2:** Fluency - Adding and subtracting through zero with number lines and in context.

✱ **Required:** 7/10    ✱ **Student Navigation:** on
✱ **Randomised:** off

**11.** 8 + -5 =

 ▪ **3**

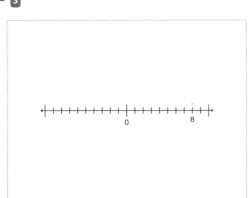

**12.** Calculate 7 - -2.

 ▪ **9**

**13.** 12 + -32 =

 ▪ **-20**

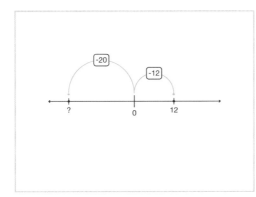

**14.** What is -17 + 22 =

 ▪ **5**

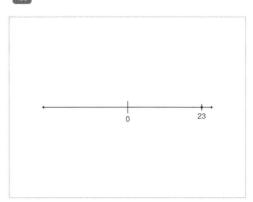

**15.** In August the temperature in Alberta, Canada is 23 °C: in January the temperature falls by 38 °C. What is the temperature in Alberta in January?
*Don't include the units in your answer.*

▪ **-15**

**16.** Select the symbol that makes the statement true.
    -14 + -12 ___ 36 - 62

▪ < ▪ **=** ▪ >

1/3

**Level 2:** *cont.*

**17.** What number is missing from the following calculation?

**1**
**2**
**3**

9 - 17 - _ = -5.

■ -3

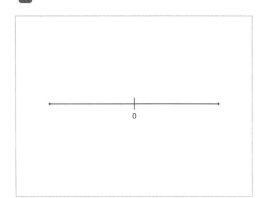

**18.** What is -18 - 25?

**1**
**2**
**3**

■ -43

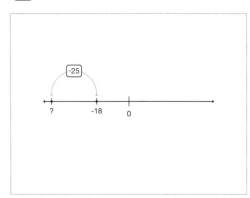

**19.** When you add 11 to -7, what number do you get?

**1**
**2**
**3**

■ 4

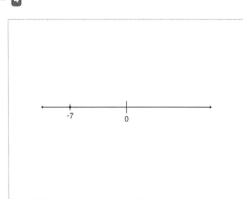

**20.** Tobias is playing a computer game and has a score of 8. He makes a mistake and loses 45 points. What is his new score?

**1**
**2**
**3**

■ -37

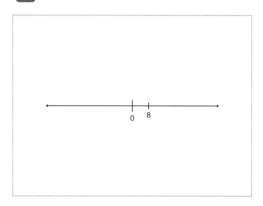

**Level 3:** Reasoning - Using understanding of negative numbers in varying contexts.

✿ **Required:** 5/5     ✿ **Student Navigation:** on
✿ **Randomised:** off

**21.** The following calculation needs to contain brackets to make it true:

☐
☒
☐

-18 + 3 x 4 = -6.
Select the correct calculation.

1/3

■ -18 + (3 x 4) = -6   ■ (-18 + 3) x 4 = -6   ■ (-18) + 3 x 4 = -6

**22.** Fleur says, "If I have -16 and add 14 to it then I will get -30 as my answer".

**a**
**b**
**c**

Is Fleur correct? Explain your answer.

**23.** In metres, what is the difference between the highest point above sea level and lowest point below sea level as shown by the table?

**a**
**b**
**c**

*Include the units m (metres) in your answer.*

■ 9,014 m   ■ 9014 m   ■ 9014 metres   ■ 9,014 metres

| country | highest point (m) | lowest point (m) |
|---|---|---|
| Bermuda | 76 | 0 |
| Egypt | 2,629 | -133 |
| India | 8,586 | -2 |
| Jordan | 1,854 | -428 |
| Switzerland | 4,634 | 198 |

**24.** Select the 2 numbers with a difference of 32.

☐
☒
☐

■ -13.4   ■ 16.2   ■ -12.6   ■ 15.8   ■ 13.4   ■ -15.8

2/6

**Level 3:** *cont.*

25. Suzie is drawing a square on the axes shown.
    **a b c** What are the co-ordinates of the square's final point?

    ▪ (2, -1)

    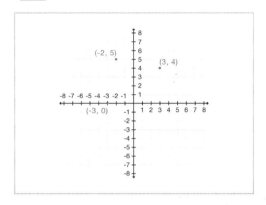

**Level 4:** Problem Solving - Adding and subtracting negative numbers in real-life contexts.

✵ **Required:** 5/5    ✵ **Student Navigation:** on
✵ **Randomised:** off

26. Add two-side-by-side numbers to find the answer
    **1 2 3** to the box above. What number is missing from the striped circle?

    ▪ 94

    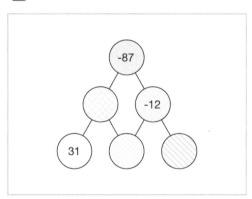

27. Use each card once to make the smallest possible
    **a b c** 4-digit number.
    What do you need add to this number to equal -8,000?

    ▪ -569

    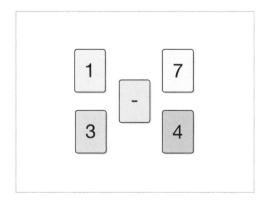

28. Jacob lives in Helsinki and his friend Callum lives in
    **a b c** Mexico City. When Jacob spoke to Callum, he said it was 47° C in Mexico city, which was 53.5° C warmer than in Helsinki.
    At what time did Jacob speak to Callum?
    Give your answer in the form **18:20** (24-hr clock)

    ▪ 21.50   ▪ 21:50

    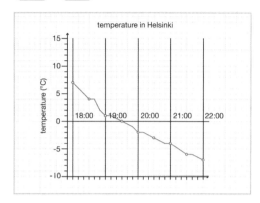

29. What is the value of the square?
    **1 2 3** ▪ -10

    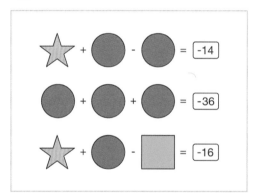

30. Morgan writes down a number sequence where
    **1 2 3** the numbers increase in order of size.
    The 5th number is 37 greater than the 2nd number. What is the 3rd number in Morgan's sequence?
    The range is 52.
    The numbers have a sum of 0.
    The 3rd and 4th numbers have a difference of 1.

    ▪ 2

    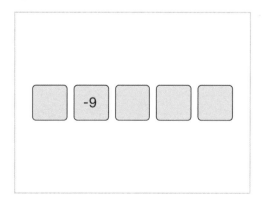

# Mathematics Y7

## Multiplication and Division

Inverse Operations

Written Methods Division

Written Methods Multiplication

Negative Numbers

Powers of 10

Decimals

# Recognise and use relationships between multiplication and division

**Competency:**    Recognise and use relationships between operations including inverse operations.

**Quick Search Ref:**    10259

**Correct:** Correct.    **Wrong:** Incorrect, try again.    **Open:** Thank you.

**Level 1:**    Reasoning - Recognise and use relationships between multiplication and division including inverse operations.

✼ **Required:** 5/5    ✼ **Student Navigation:** on    ✼ **Randomised:** off

**1.** Which of the following calculations are **incorrect**?

■ 5 x 6 = 6 x 5   ■ 40 ÷ 8 = 8 ÷ 40   ■ 9 + 17 = 17 + 9

■ 76 - 43 = 43 - 76

2/4

**2.** Explain why 4 + 3 x 10 x 2 gives the same answer as 2 x 3 x 10 + 4.

**3.** Using the fact 144 ÷ 6 = 24, which of the following calculations are true?
*There are 3 correct answers*

3/6   ■ 24 x 6 = 144   ■ 6 ÷ 144 = 24   ■ 6 x 24 = 144
■ 144 x 24 = 6   ■ 144 x 6 = 24   ■ 144 ÷ 24 = 6

**4.** Hammera says, '450 ÷ 18 is the same as 450 ÷ 3 ÷ 6'. Is she correct? Explain your answer.

**5.** Which two equations help you fill in the blank in the following equation?
___ ÷ 7 = 5.

2/4   ■ 5 ÷ 7 = ___   ■ 5 x 7 = _____   ■ 7 ÷ 5 = ___   ■ 7 x 5 = _____

**Level 2:**    Problem Solving - Recognise and use relationships between multiplication and division including inverse operations.

✼ **Required:** 5/5    ✼ **Student Navigation:** on
✼ **Randomised:** off

**6.** The keys 4 and 9 are broken on George's calculator. He wants to work out 49 × 5. What multiplication could he do on his calculator instead?

■ 7 * 5 * 7   ■ 7 x 5 x 7   ■ 7 * 7 * 5   ■ 7 x 7 x 5
■ 5 x 7 x 7   ■ 5 * 7 * 7

**7.** On Monday, Karen shares 48 paperclips with her group of friends.
On Tuesday she shares 48 paperclips with just half her friends.
On Tuesday each friend receives _____ the amount of paperclips they received on Monday. What word fills the blank?

■ Twice   ■ Double

**8.** To divide a number by 4, you half the number 2 times because 4 = 2 x 2.
How many times would you half a number to divide it by 32?

■ 5

**9.** Which of the following sets of calculations give the same answer?

■ (6 + 3) x 4 and 6 x (3 + 4)   ■ 2 x (4 + 7) and (2 x 4) + (2 x 7)

1/3   ■ 17 - 4 + 5 and 4 + 5 + 17

**10.** Jane made a mistake when performing a calculation and multiplied by 54 instead of 45. Her answer of 1,188 was 198 too large. What number did she multiply 54 by?

■ 22

# Divide integers by integers (bus stop method)

**Competency:** Use the four operations, including formal written methods, applied to integers, decimals, proper and improper fractions and mixed numbers, all both positive and negative.

**Quick Search Ref:** 10206

Correct: Correct.    Wrong: Incorrect, try again.    Open: Thank you.

**Level 1:** Understanding - Divide integers with divisors up to 20 with integer and decimal answers.

✱ **Required:** 7/10    ✱ **Student Navigation:** on    ✱ **Randomised:** off

**1.** Select the term that best describes division.

- Product ■ Lots of ■ Share equally ■ Sum
- Difference

1/5

**2.** Given that 23 x 6 = 138, which of the following are correct.
*Select three correct answers.*

3/6  ■ 138 ÷ 23 = 6  ■ 6 x 23 = 138  ■ 6 ÷ 138 = 23
■ 6 ÷ 23 = 138  ■ 138 ÷ 6 = 23  ■ 23 ÷ 6 = 138

**3.** What is 9,936 ÷ 12?

■ 828

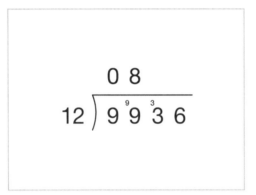

**4.** Calculate 10,556 ÷ 14.

■ 754

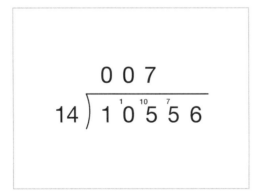

**5.** What is 9 ÷ 8 as a decimal?

 ■ 1.125

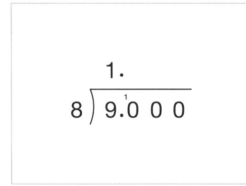

**6.** Divide 642 by 15.

 ■ 42.8

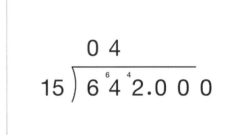

**7.** What is 9,354 ÷ 12?

 ■ 779.5

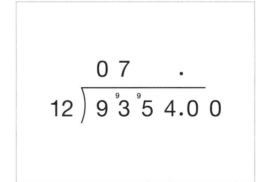

**8.** Which division gives the **smallest** answer?

 ■ 306 ÷ 18  ■ 285 ÷ 15  ■ 285 ÷ 19

1/3

**Level 1: cont.**

**9.** How many 20s are there in 4,089?

 ▪ 204.45

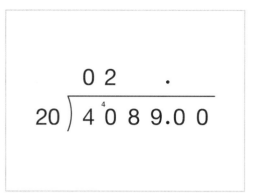

**10.** Complete this calculation: 46,092 ÷ 16 = ____ .

▪ 2,880.75  ▪ 2880.75

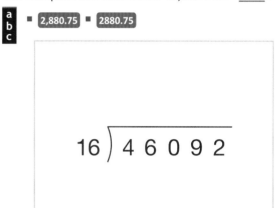

**Level 2:** Fluency - Divide larger integers with divisors up to 50 with decimal answers and answer questions in context.

✻ **Required:** 7/10    ✻ **Student Navigation:** on
✻ **Randomised:** off

**11.** Fill in the missing number:
522 ÷ ___ = 29.

▪ 18

**12.** Calculate 1,236 ÷ 48.

▪ 25.75

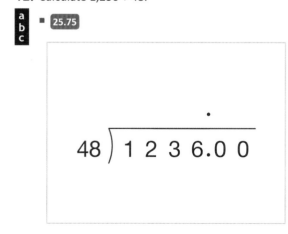

**13.** Work out the answer to 19,551 ÷ 40.

▪ 488.775

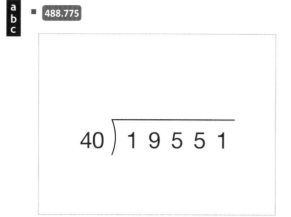

**14.** Calculate 45,623 ÷ 50.

▪ 912.46

**15.** Calculate 1,452 ÷ 41.
*Round your answer to 2 decimal places.*

▪ 35.41

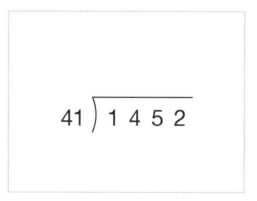

**16.** To raise money for charity, a sports club is doing a relay race of 225 kilometres (km). Each runner will run 9 km.
How many runners are needed to complete the race?

▪ 25

**17.** If a rope is 624 centimetres (cm) long and it is cut into 24 equal length pieces, how long is each new piece of rope?
*Include the units cm (centimetres) in your answer.*

▪ 26 centimetres  ▪ 26 cm

**Level 2: cont.**

**18.** What is 428 ÷ 16?

a
b
c   ▪ 26.75

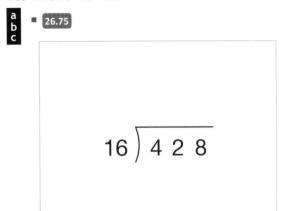

**19.** A cake of mass 2 kilograms (kg) is cut into 16 equal
a   pieces. What is the mass of one piece of cake?
b   *Include the units kg (kilograms) in your answer.*
c

▪ 0.125 kilograms  ▪ 0.125 kg

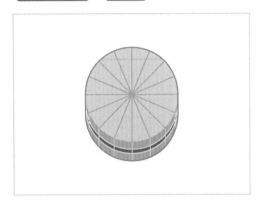

**20.** A company has 290 boxes that need to be
a   delivered. One van holds 13 boxes.
b   How many vans will be required?
c

▪ 23

**Level 3:** Reasoning - Divide integers by integers (bus
stop method).

✿ **Required:** 5/5   ✿ **Student Navigation:** on
✿ **Randomised:** off

**21.** 4,500 people travel in 55-seater coaches from
a   Berlin to Venice.
b   Sofia thinks they will need 81 coaches.
c   Is she correct? Explain your answer.

**22.** Work out the answer to 2 ÷ 11. What do you
a   notice?
b
c

**23.** Calculate the 5th term in the sequence shown.

a
b   ▪ 0.15625
c

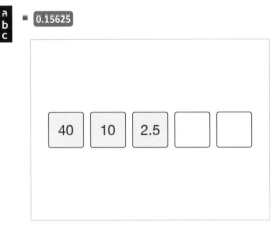

**24.** Parminder has calculated 97 ÷ 8 = 11.
a   Explain what he has done wrong.
b
c

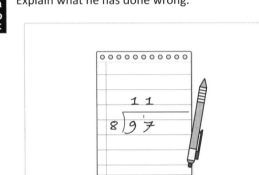

**25.** Henry works out the answer to 9,441 ÷ 18 as 524.9
a   but Abigail thinks he is wrong.
b   What is the correct answer?
c

▪ 524.5

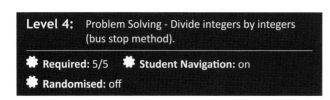

**Level 4:** Problem Solving - Divide integers by integers (bus stop method).

✱ **Required:** 5/5   ✱ **Student Navigation:** on
✱ **Randomised:** off

**26.** A parallelogram has an area of 216 square
**a** centimetres (cm²). The base of the shape is 0.45
**b** metres (m). Calculate the perpendicular height of
**c** the shape.
*Include the units cm (centimetres) in your answer.*

▪ 4.8 cm  ▪ 4.8 centimetres

area = 216 cm²

0.45 m

**27.** If it takes Natalie three hours to read 48 pages of
**a** her book, how long does it take her to read one
**b** page in *seconds*?
**c** *Include the units s (seconds) in your answer.*

▪ 225 seconds  ▪ 225 s

**28.** Jonathan and Rabiya win a competition and share
**a** the prize money. Jonathan gets three times as
**b** much as Rabiya. The total prize money is £330.
**c** How much does Jonathan receive?
*Include the £ sign in your answer.*

▪ £247.50

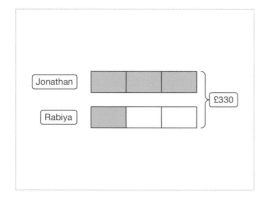

Jonathan

Rabiya

£330

**29.** Fill in the four missing digits. Using the four
**a** missing digits, what is the **largest number** you can
**b** make?
**c**

▪ 7221  ▪ 7,221

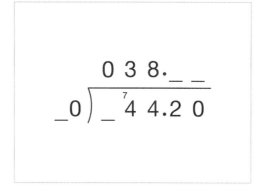

$$0\ 3\ 8.\ \_\ \_$$
$$\_\ 0\ )\ \_\ \overset{7}{4}\ 4.2\ 0$$

**30.** Complete the puzzle. What is the **smallest** 5-digit
**a** number that you can make from the digits in
**b** the shaded circles?
**c**

▪ 10269  ▪ 10,269

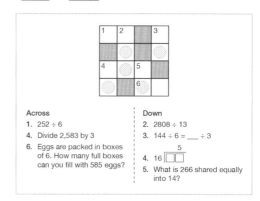

**Across**
1. 252 ÷ 6
4. Divide 2,583 by 3
6. Eggs are packed in boxes of 6. How many full boxes can you fill with 585 eggs?

**Down**
2. 2808 ÷ 13
3. 144 ÷ 6 = ___ ÷ 3
4. 16 [  ][  ]
5. What is 266 shared equally into 14?

# Multiply and divide by positive and negative powers of ten

**Competency:** Use the four operations, including formal written methods, applied to integers, decimals, proper and improper fractions and mixed numbers, all both positive and negative.

**Quick Search Ref:** 10239

Correct: Correct.    Wrong: Incorrect, try again.    Open: Thank you.

**Level 1:** Understanding - Multiply and divide integers and decimals up to 1 decimal place by positive and negative powers of ten.

✱ Required: 7/10        ✱ Student Navigation: on        ✱ Randomised: off

---

**1.** What is 3.8 x 1,000?

a b c  ▪ 3800 ▪ 3,800

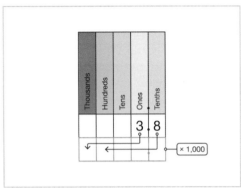

**2.** What is 6.2 ÷ 1,000?

a b c  ▪ 0.0062

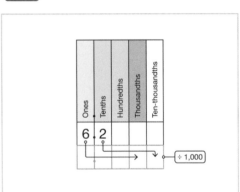

**3.** Give 0.1 as a fraction.

a b c  ▪ 1/10

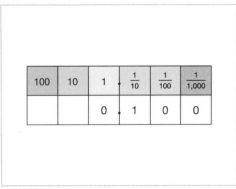

**4.** What letter represents the calculation which means the same as ÷ 0.1 (divided by 0.1)?

a b c  ▪ C

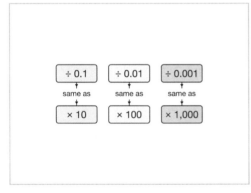

**5.** What is 3.5 ÷ 0.1?

a b c  ▪ 35

| ÷ 0.1 | ÷ 0.01 | ÷ 0.001 |
|---|---|---|
| same as | same as | same as |
| × 10 | × 100 | × 1,000 |

**6.** What letter represents the calculation which means the same as x 0.01 (multiply by 0.01)?

a b c  ▪ L

| A | × 1,000 | G | ÷ 10 |
|---|---|---|---|
| B | × 0.1 | H | ÷ 0.01 |
| C | × 10 | I | ÷ 0.1 |
| D | × 100 | J | ÷ 0.001 |
| E | × 0.001 | K | ÷ 1,000 |
| F | × 0.01 | L | ÷ 100 |

**Level 1:** *cont.*

**7.** Calculate 475.5 x 0.01.

 ▪ 4.755

| 100 | 10 | 1 | $\frac{1}{10}$ | $\frac{1}{100}$ | $\frac{1}{1,000}$ | $\frac{1}{10,000}$ |
|---|---|---|---|---|---|---|
| 4 | 7 | 5 | 5 | 0 | 0 | 0 |
| | | | | | | |

**8.** What letter represents the calculation which means the same as x 0.001 (multiply by 0.001)?

 ▪ K

| A | × 1,000 | G | ÷ 10 |
|---|---|---|---|
| B | × 0.1 | H | ÷ 0.01 |
| C | × 10 | I | ÷ 0.1 |
| D | × 100 | J | ÷ 0.001 |
| E | × 0.001 | K | ÷ 1,000 |
| F | × 0.01 | L | ÷ 100 |

**9.** Calculate 432.6 × 0.001.

 ▪ 0.4326

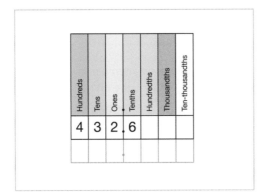

| Hundreds | Tens | Ones | Tenths | Hundredths | Thousandths | Ten-thousandths |
|---|---|---|---|---|---|---|
| 4 | 3 | 2 | 6 | | | |
| | | | | | | |

**10.** What is 9.2 ÷ 0.1?

 ▪ 92

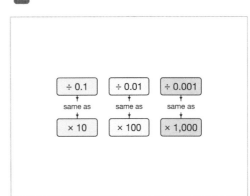

| ÷ 0.1 | ÷ 0.01 | ÷ 0.001 |
|---|---|---|
| same as | same as | same as |
| × 10 | × 100 | × 1,000 |

**Level 2:** Fluency – Multiply and divide decimals up to 4 d.p. by positive and negative powers of ten.

✱ **Required:** 7/10    ✱ **Student Navigation:** on
✱ **Randomised:** off

**11.** What number is 1,000 times bigger than 109.85?

 ▪ 109850 ▪ 109,850

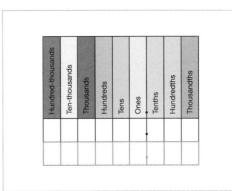

| Hundred-thousands | Ten-thousands | Thousands | Hundreds | Tens | Ones | Tenths | Hundredths | Thousandths |
|---|---|---|---|---|---|---|---|---|
| | | | | | | | | |
| | | | | | | | | |

**12.** What number is 10 times smaller than 54.708?

 ▪ 5.4708

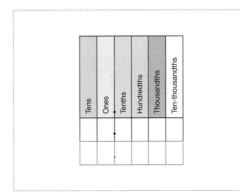

| Tens | Ones | Tenths | Hundredths | Thousandths | Ten-thousandths |
|---|---|---|---|---|---|
| | | | | | |
| | | | | | |

**Level 2:** *cont.*

**13.** Which operation can you perform on the top number to make it equal to the bottom number? *Select the two correct answers.*

2/7 ▪ ÷ 1,000 ▪ ÷ 0.001 ▪ x 0.001 ▪ x 100 ▪ x 1,000
▪ ÷ 0.01 ▪ x 0.01

| 100 | 10 | 1 | . | $\frac{1}{10}$ | $\frac{1}{100}$ | $\frac{1}{1,000}$ |
|-----|----|----|---|------|-------|--------|
|     |    | 0  | . | 0    | 9     | 5      |
|     | 9  | 5  | . | 0    | 0     | 0      |

**14.** What is 0.785 × 0.01?

▪ 0.00785

**15.** Choose the two calculations that make the following calculation true:

2.74 _____ = 274

2/7 ▪ x 100 ▪ ÷ 0.001 ▪ x 0.001 ▪ x 1000 ▪ x 0.1 ▪ ÷ 0.01
▪ x 0.01

**16.** Convert 4.56 kilometres to metres (m).
*Include the units m (metres) in your answer.*

▪ 4,560 metres ▪ 4,560 m ▪ 4560 metres ▪ 4560 m

**17.** What is the missing number?

_____ ÷ 0.001 = 45.68.

▪ 0.04568

**18.** Find the value of each calculation and arrange them in ascending order (smallest first).

▪ 24.37 x 0.001 ▪ 243.7 ÷ 1,000 ▪ 2.437 x 10
▪ 2437 x 0.1 ▪ 24.37 ÷ 0.01

**19.** Write 43.95% as a decimal.

▪ 0.4395

**20.** Which calculation can you perform on the top number to make it equal to the bottom number? *Select the two correct answers.*

2/7 ▪ ÷ 1000 ▪ ÷ 0.001 ▪ x 0.001 ▪ ÷ 100 ▪ x 1000
▪ ÷ 0.01 ▪ x 0.01

| 100 | 10 | 1 | . | $\frac{1}{10}$ | $\frac{1}{100}$ | $\frac{1}{1,000}$ | $\frac{1}{10,000}$ |
|-----|----|----|---|------|-------|--------|---------|
|     | 7  | 8  | . | 5    | 3     | 0      | 0       |
|     |    | 0  | . | 7    | 8     | 5      | 3       |

**Level 3:** Reasoning - Multiply and divide any number by positive and negative powers of ten.

✱ **Required:** 5/5   ✱ **Student Navigation:** on
✱ **Randomised:** off

**21.** Explain why x 0.01 is the same as ÷ 100.

**22.** What is the missing number?

4 x 0.01 x _____ = 2.4.

▪ 60

**23.** Diego performs the following calculation:
0.708 ÷ 10 = 0.78.
Explain what he has done wrong.

**24.** Work out the missing number in each of the calculations. Which is the odd one out?

▪ 4.7 x _____ = 0.047 ▪ 39 ÷ _____ = 0.39
1/3 ▪ 0.0017 ÷ ___ = 0.17

**25.** Gary says he's moved all of the digits three places to the left because his teacher asked him to multiply by 1,000. What has Gary done wrong?

| 100 | 10 | 1 | . | $\frac{1}{10}$ | $\frac{1}{100}$ | $\frac{1}{1,000}$ | $\frac{1}{10,000}$ |
|-----|----|----|---|------|-------|--------|---------|
|     |    | 0  | . | 3    | 7     | 0      | 0       |
|     | 3  | 7  | . | 0    | 0     | 0      | 0       |

**Level 4:** Problem Solving - Multiply and divide any number by positive and negative powers of ten.

✱ **Required:** 5/5    ✱ **Student Navigation:** on

✱ **Randomised:** off

**26.** Find the four equivalent pairs in the grid and
**a** identify the **odd one out**.
**b** Perform this calcualtion on the number 543,210.
**c** What is the answer?

■ 54,321   ■ 54321

| × 1,000 | ÷ $\frac{1}{10}$ | × 100 |
|---|---|---|
| ÷ 100 | ÷ 0.01 | ÷ 10 |
| × 10 | ÷ 0.001 | × $\frac{1}{100}$ |

**27.** 106 kilometres (km) is 5.3 centimetres (cm) on a
**a** map. What is the scale factor?
**b**
**c** ■ 2000000   ■ 2,000,000

**28.** Find the four equivalent pairs in the grid.
**1** What is the value of the odd one out?
**2**
**3** ■ 5.62

| 5.62 ÷ 10 | 562 × 0.1 | 5.62 ÷ $\frac{1}{10}$ |
|---|---|---|
| 5.62 × 1,000 | 56.2 × 10 | 562 × 0.01 |
| 0.562 ÷ 1,000 | 56.2 × $\frac{1}{100}$ | 5620 ÷ 100 |

**29.** Keisha is looking at a plan drawn to a scale of 1:50.
**1** If the area of a building on the plan is 900 cm²
**2** (square centimetres), what is the **actual area** of
**3** the building in m² (square metres)?
*Don't include the units in your answer.*

■ 225

**30.** If 82.3 is the output of the 6-step function
**1** machine, what number was the input?
**2**
**3** ■ 1.833

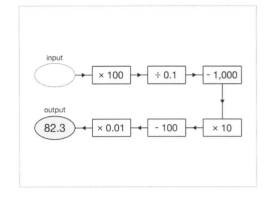

# Multiply and divide negative numbers

**Competency:** Use the four operations, including formal written methods, applied to integers, decimals, proper and improper fractions and mixed numbers, all both positive and negative.

**Quick Search Ref:** 10230

**Correct:** Correct.   **Wrong:** Incorrect, try again.   **Open:** Thank you.

**Level 1:** Understanding - The value of negative numbers and using them in simple calculations.

✹ **Required:** 7/10   ✹ **Student Navigation:** on   ✹ **Randomised:** off

**1.** What is a **negative** number?

- A number greater than zero.  ■ A number equal to zero.
- A number less than zero.

1/3

**2.** What integer makes the following equation correct?

7 + ___ = 3.

- -4

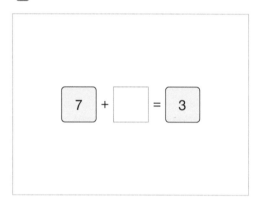

**3.** 7 x -4 means 7 lots of -4. Which calculation shows this?

- (-4) + (-4) + (-4) + (-4) + (-4) + (-4) +(-4) = -28
- 4 + 4 + 4 + 4 +4 + 4 + 4 = 28  ■ 4 - 4 - 4 - 4 - 4 - 4 - 4 = -20

1/3

**4.** Select the **two** statements that are true.

- Two like signs give a negative answer.
- Two like signs give a positive answer.
- Two unlike signs give a negative answer.
- Two unlike signs give a positive answer.

2/4

**5.** What is 8 ÷ -2?

- -4

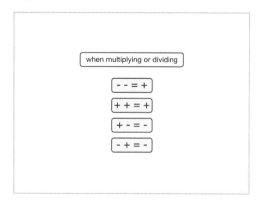

**6.** -12 x 4 = ___.

- -48

when multiplying or dividing

- - = +
+ + = +
+ - = -
- + = -

**7.** Calculate -9 ÷ -3.

- 3

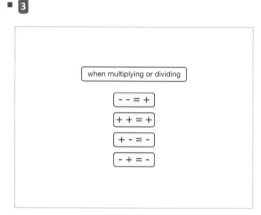

when multiplying or dividing

- - = +
+ + = +
+ - = -
- + = -

**8.** Nadia gives 3 of her pens away. What integer represents this?

- -3

**9.** -14 ÷ -7 = ___.

- 2

when multiplying or dividing

- - = +
+ + = +
+ - = -
- + = -

**10.** What is 5 x -3?

 ▪ -15

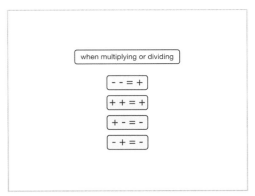

**Level 2:** Fluency - Larger numbers and calculations in context.

✱ **Required:** 7/10    ✱ **Student Navigation:** on
✱ **Randomised:** off

**11.** Calculate -24 x -130.

 ▪ 3120 ▪ 3,120

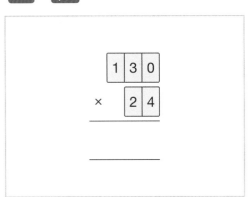

**12.** 576 ÷ -18 =

 ▪ -32

**13.** In Vladimir, Russia the average temperature in December is -3 °C. In January it gets 4 times colder.
What is the average temperature in January?
*Don't include the units in your answer.*

▪ -12

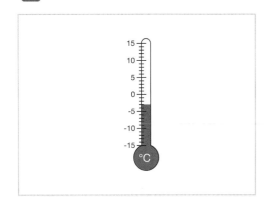

**14.** Select the two numbers that divide to make -7.

 ▪ 77 ▪ -49 ▪ 11 ▪ 63 ▪ -7 ▪ -9

2/6

**15.** What is -9²?

 ▪ 81

**16.** Select the symbol that makes the statement true.
-12 x 4 _ -45 ÷ -5

 ▪ < ▪ = ▪ >

1/3

**17.** If the top number is the product of the bottom two numbers, what number is missing?

 ▪ -8

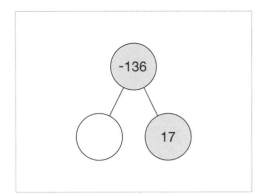

**Level 2:** *cont.*

**18.** What is -285 ÷ 19?

 ■ -15

19)285

**19.** Select the symbol that makes the statement true.
-128 ÷ -4 _ 4 x -8

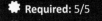 ■ < ■ = ■ >

1/3

**20.** Select the two numbers that have a product of 36.

■ 6 ■ -9 ■ -6 ■ -2 ■ 4 ■ -18

2/6

**Level 3:** Reasoning - With negative numbers including more than 2 signs.

✱ **Required:** 5/5   ✱ **Student Navigation:** on
✱ **Randomised:** off

**21.** The top number is the product of the bottom 2.
 The numbers must be integers. How many possible combinations are there?

■ 6

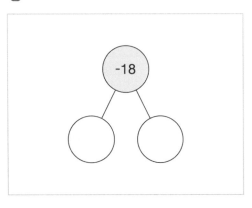

-18

**22.** -4 ÷ -2 x -7 = ___.

■ -14

**23.** The square root of 16 has 2 values, 4 and -4.
Explain why.

**24.** What is -5³?

 ■ -125

**25.** Melissa is thinking of a number. She multiplies it by 0, will her answer be positive or negative? Explain your answer.

**Level 4:** Problem Solving - With negative numbers including algebra, inverse calculations and unknown values.

✱ **Required:** 5/5   ✱ **Student Navigation:** on
✱ **Randomised:** off

**26.** -2(5x - 4) = 24
What is the value of *x*?

■ -2

**27.** Elle is thinking of a number.
She multiplies it by 9 and then subtracts -50.
She adds -2 and finally divides by -3 to get the answer 5.
What number did she start with?

■ -7

**28.** What is the value of the rectangle?

■ -5

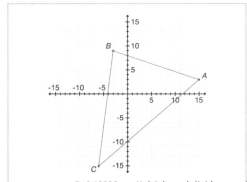

(a) ★ × ▲ = -24

(b) ▲ - ★ = -11

(c) ▲ × ★ × ▬ = 120

**29.** Jamil has drawn a triangle on the axes. He wants to draw a new shape that has the same origin (0,0) but is 3 times smaller than the original shape. What will the co-ordinates be for point *B* of the new shape?

■ (-1, 3) ■ (-1,3)

**30.** By using all 5 number cards and the 4 operation
 cards what is the biggest number you can make?
You must use each of the cards only once.

■ 220

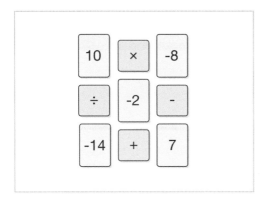

## Practise Four Operations with Negative Numbers

**Competency:**    Use the four operations with negative numbers

**Quick Search Ref:**    10529

Correct: Correct.    Wrong: Incorrect. Try again.    Open: Thank you.

**Level 1:**    Use the four operations with negative numbers.

✿ **Required:** 20/48    ✿ **Student Navigation:** off    ✿ **Randomised:** on

**1.**  -7 + 3 =

[1/2/3]  ▪ -4

**2.**  -8 + 6 =

[1/2/3]  ▪ -2

**3.**  -15 + 9 =

[1/2/3]  ▪ -6

**4.**  -13 + 21 =

[1/2/3]  ▪ 8

**5.**  -8 + 13 =

[1/2/3]  ▪ 5

**6.**  -27 + 36 =

[1/2/3]  ▪ 9

**7.**  9 + -17 =

[1/2/3]  ▪ -8

**8.**  15 + -22 =

[1/2/3]  ▪ -7

**9.**  12 + -15 =

[1/2/3]  ▪ -3

**10.**  19 + -12 =

[1/2/3]  ▪ 7

**11.**  27 + -19 =

[1/2/3]  ▪ 8

**12.**  14 + -8 =

[a/b/c]  ▪ 6

**13.**  -9 + -7 =

[1/2/3]  ▪ -16

**14.**  -17 + - 15 =

[1/2/3]  ▪ -32

**15.**  -27 + -35 =

[1/2/3]  ▪ -62

**16.**  7 - 11 =

[1/2/3]  ▪ -4

**17.**  15 - 22 =

[1/2/3]  ▪ -7

**18.**  18 - 24 =

[1/2/3]  ▪ -6

**19.**  -14 - 6 =

[1/2/3]  ▪ -20

**20.**  -24 - 15 =

[1/2/3]  ▪ -39

**Level 1: *cont.***

**21.** -26 - 16 =

■ -42

**22.** 9 - -6 =

■ 15

**23.** 18 - -15 =

■ 33

**24.** 23 - -31 =

■ 54

**25.** -7 - -4 =

■ -3

**26.** -16 - -9 =

■ -7

**27.** -27 - -18 =

■ -9

**28.** -6 - -10 =

■ 4

**29.** -13 - -17 =

■ 4

**30.** -23 - -34 =

■ 11

**31.** 8 × -4 =

■ -32

**32.** 9 × -3 =

■ -27

**33.** 11 × -5 =

■ -55

**34.** -5 × 5 =

■ -25

**35.** -6 × 2 =

■ -12

**36.** -8 × 3 =

■ -24

**37.** -3 × -9 =

■ 27

**38.** -8 × -7 =

■ 56

**39.** -12 × -9 =

■ 108

**40.** 48 ÷ -6 =

■ -8

**41.** 36 ÷ -9 =

■ -4

**42.** 21 ÷ -3 =

■ -7

**43.** -32 ÷ 4 =

■ -8

**44.** -54 ÷ 6 =

■ -9

## Level 1: *cont.*

**45.** -81 ÷ 9 =
1
2
3  ▪ -9

**46.** -48 ÷ -12 =
1
2
3  ▪ 4

**47.** -27 ÷ -9 =
1
2
3  ▪ 3

**48.** -77 ÷ -7 =
1
2
3  ▪ 11

## Level 2: Find the missing value using the four operations with negative numbers.

✳ **Required:** 8/16    ✳ **Student Navigation:** off
✳ **Randomised:** on

**49.** -12 + __ = -5
1
2
3  ▪ 7

**50.** -24 + __ = 18
1
2
3  ▪ 42

**51.** 23 + __ = 8
1
2
3  ▪ -15

**52.** 17 + __ = -3
1
2
3  ▪ -20

**53.** -25 + __ = -37
1
2
3  ▪ -12

**54.** 27 - __ = -4
1
2
3  ▪ 31

**55.** -6 - __ = -15
1
2
3  ▪ 9

**56.** 13 - __ = 36
1
2
3  ▪ -23

**57.** -23 - __ = -6
1
2
3  ▪ -17

**58.** -19 - __ = 5
1
2
3  ▪ -24

**59.** 6 × __ = -42
1
2
3  ▪ -7

**60.** -56 ÷ __ = 8
1
2
3  ▪ -7

**61.** -6 × __ = -24
1
2
3  ▪ 4

**62.** -8 × __ = 72
1
2
3  ▪ -9

**63.** 24 ÷ __ = -3
1
2
3  ▪ -8

**64.** -64 ÷ __ = -8
1
2
3  ▪ 8

# Divide decimals by integers (bus stop method)

**Competency:** Use the four operations, including formal written methods, applied to integers, decimals, proper and improper fractions and mixed numbers, all both positive and negative.

**Quick Search Ref:** 10257

**Correct:** Correct **Wrong:** Incorrect, try again. **Open:** Thank you.

**Level 1:** Understanding - Dividing numbers up to 3 decimal places by divisors up to 12.

❋ **Required:** 9/10    ❋ **Student Navigation:** on    ❋ **Randomised:** off

---

**1.** Calculate 7,623 ÷ 12.

 ▪ 635.25

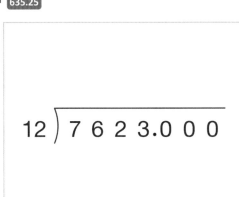

**2.** Use the bus stop method to divide 1.2 by 4.

 ▪ 0.3

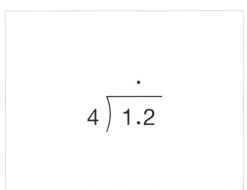

**3.** Work out the answer to 0.224 ÷ 7.

 ▪ 0.032

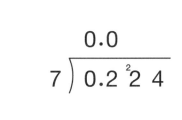

**4.** What is 0.812 ÷ 4?

 ▪ 0.203

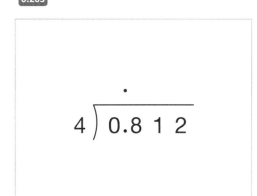

**5.** Work out the answer to 228.6 ÷ 9.

▪ 25.4

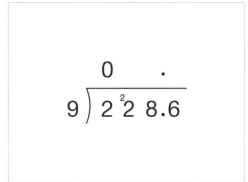

**6.** Split 158.643 into 3 equal parts.

▪ 52.881

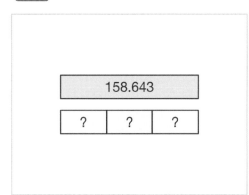

**Level 1:** *cont.*

**7.** What is 5,527.74 shared equally by 6?

a b c  ▪ 921.29

**8.** Calculate 6,204.165 ÷ 11.

a b c  ▪ 564.015

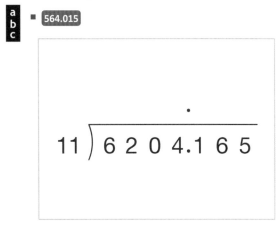

**9.** Share 714.84 by 12.

a b c  ▪ 59.57

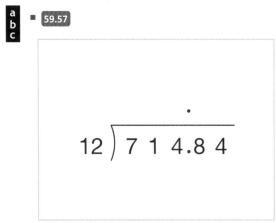

**10.** Calculate 24.896 ÷ 8 = _____.

a b c  ▪ 3.112

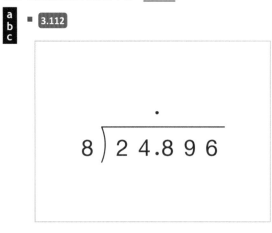

**Level 2:**  Fluency - Divide numbers up to 4 decimal places by divisors up to 50 (including questions in context).

✱ **Required:** 8/10    ✱ **Student Navigation:** on
✱ **Randomised:** off

**11.** Divide 3,327.1 by 14.

1 2 3  ▪ 237.65

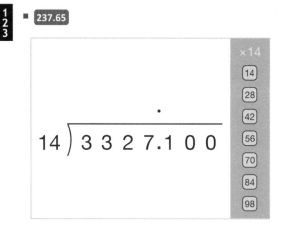

**12.** Complete this calculation:

a b c  _____ x 23 = 57,740.81

▪ 2,510.47  ▪ 2510.47

**13.** 18.5688 ÷ 6 = _____ x 4.

a b c  ▪ 0.7737

**14.** Select the symbol that makes the following statement true:

123.75 ÷ 45 _____ 118.234 ÷ 31.

1/3   ▪ =  ▪ <  ▪ >

**15.** What is 4694.783 ÷ 23?

a b c  ▪ 204.121

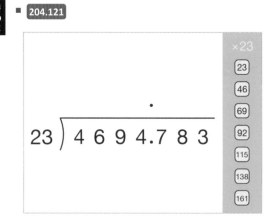

**16.** Select the correct calculation for the following problem.

Richard has 5.8 kilograms (kg) of coffee that he wants to split into 0.25 kg bags. How many bags will he need?

1/6

▪ 5.8 x 0.25  ▪ 0.25 ÷ 5.8  ▪ 0.25 x 5.8  ▪ 5.8 ÷ 0.25

▪ 5.8 + 0.25  ▪ 5.8 - 0.25

**Level 2: cont.**

**17.** There are 735 people in the queue for a roller coaster. A roller coaster car seats 14 people for each ride.
a
b
c
How many rides will there be before everyone has had a go?

■ 53

**18.** Calculate 2,503.056 ÷ 48.
a
b
c
■ 52.147

**19.** Heidi bought a set of eight plates costing total of £77.20. How much did each plate cost?
a
b
c
*Include the £ sign in your answer.*

■ £9.65

**20.** Suzie bought 4 rolls of paper towels, with a total length of 87.88 metres (m). How many metres were on each roll?
a
b
c
*Include the units m (metres) in your answer.*

■ 21.97 m ■ 21.97 metres

**Level 3:** Reasoning - Divide decimals by integers (bus stop method).
━━━━━━━━━━━━━━━━━━━━━━━━━━━━━
✹ **Required:** 5/5   ✹ **Student Navigation:** on
✹ **Randomised:** off

**21.** Farooq calculates 2210.496 ÷ 24 to equal 92.14.
a
b
c
Sabrina get the answer 92.104.
Who is correct? Explain what the other person has done wrong.

**22.** Explain what what mistake has been made in the division shown.
a
b
c

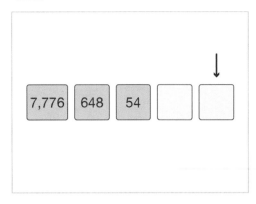

**23.** Work out the 5th term of the sequence.
a
b
c
7,776; 648; 54; _____; _____

■ 0.375

┌─────────────────────────────────────┐
│                              ↓        │
│   ┌──────┐ ┌────┐ ┌────┐ ┌────┐ ┌────┐│
│   │7,776 │ │648 │ │ 54 │ │    │ │    ││
│   └──────┘ └────┘ └────┘ └────┘ └────┘│
│                                        │
└─────────────────────────────────────┘

**24.** Select the symbol that makes the following statement true:
☐
☒
☐
88.02 ÷ 27 ____ 68.25 ÷ 21

1/3   ■ < ■ > ■ =

**25.** Is it **always**, **sometimes** or **never** true that when you divide a decimal, you will get a decimal answer.
☐
☒
☐
1/3   ■ Always ■ Sometimes ■ Never

**Level 4:** Problem Solving - Divide decimals by integers (bus stop method).
━━━━━━━━━━━━━━━━━━━━━━━━━━━━━
✹ **Required:** 5/5   ✹ **Student Navigation:** on
✹ **Randomised:** off

**26.** A sheet of labels costs £0.85. There are 40 labels on each sheet.
a
b
c
What is the cost of one label to the nearest whole penny?
*Include the p (pence) symbol in your answer.*

■ 2 pence ■ 2 p

**27.** Melissa thinks of a number and adds 7 to it. She then multiplies the total by 16 to get an answer of 152.16.
1
2
3
What number is Melissa thinking of?

■ 2.51

**Level 4:** *cont.*

**28.** Using each of the digit cards once, make the
**a**   division calculation that has the closest possible
**b**   answer to 250.
**c**   What is the answer to your calculation?

▪ 248.6

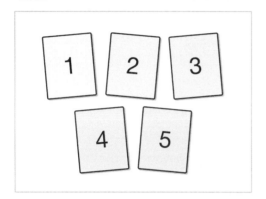

**29.** In the pyramid, the product of two side-by-side
**a**   bricks is the answer to the box above.
**b**   What number is missing from box marked $x$?
**c**

▪ 0.30625

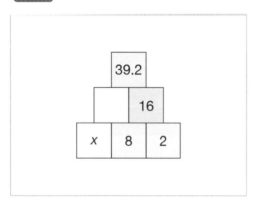

**30.** The perimeter of the shape is 45 metres (m) and
**a**   the area is 55.75 m².
**b**   What is the length of the side marked $y$?
**c**   Include the units *m in your answer.*

▪ 2.75 metres   ▪ 2.75 m

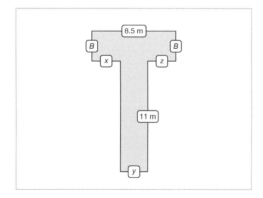

# Multiply integers and decimals (column method)

**Competency:** Use the four operations, including formal written methods, applied to integers, decimals, proper and improper fractions and mixed numbers, all both positive and negative

**Quick Search Ref:** 10093

Correct: Correct.     Wrong: Incorrect, try again.     Open: Thank you.

**Level 1:** Understanding - Multiply up to 4 digits by 2-digit integers and decimals up to 2 decimal places.

🌸 **Required:** 9/10          🌸 **Student Navigation:** on          🌸 **Randomised:** off

---

**1.** When you complete the first part of the calculation, what number will go in the highlighted box?

a b c

▪ 5,096  ▪ 5096

**2.** When you complete the second part of the calculation, what number will go in the highlighted box?

a b c

▪ 25,480  ▪ 25480

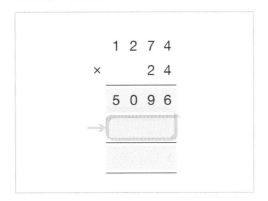

**3.** What is the answer to 1,274 x 24?

a b c

▪ 30576  ▪ 30,576

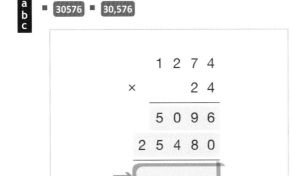

**4.** Calculate 1.2 x 0.5.

a b c

▪ 0.6

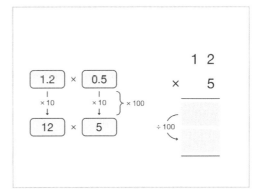

**5.** 3.23 x ___ = 323.

1 2 3

▪ 100

**6.** Work out the answer to 0.27 x 1.4.

a b c

▪ 0.378

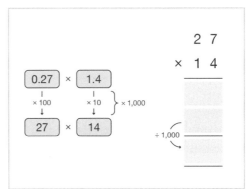

**7.** Select the calculations that will have 2 numbers after the decimal point in the answer.
*There are three correct answers.*

3/5

▪ 1.3 x 2.4  ▪ 6.14 x 1.3  ▪ 2.48 x 96.14  ▪ 8.9 x 0.1
▪ 5.4 x 7.2

---

**Level 1:** *cont.*

**8.** Use the column method to calculate 451 x 62.

a b c ▪ 27962 ▪ 27,962

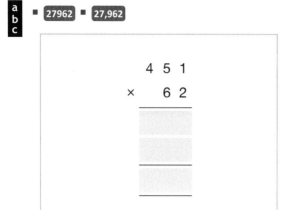

```
      4 5 1
   ×   6 2
   _____
```

**9.** Calculate 12.7 x 1.5.

1 2 3 ▪ 19.05

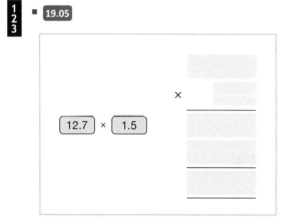

**10.** Choose the correct answer to 3.23 x 1.7.

☐☒☐ ▪ 549.1 ▪ 5.491 ▪ 5491 ▪ 54.91

1/4

**Level 2:** Fluency - Multiply integers with up to 4 digits by a 3-digit integer and decimals with up to 4 decimal places.

❋ **Required:** 7/10   ❋ **Student Navigation:** on
❋ **Randomised:** off

**11.** What is 124 x 325?

a b c ▪ 40300 ▪ 40,300

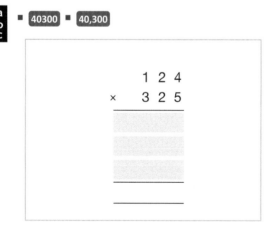

```
      1 2 4
   ×  3 2 5
   _____
```

**12.** Calculate 1,495 x 263.

a b c ▪ 393185 ▪ 393,185

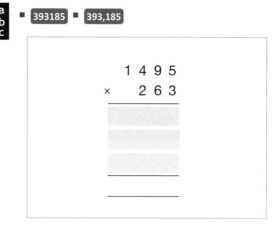

```
      1 4 9 5
   ×    2 6 3
   _____
```

**13.** What is the **product** of 289 and 0.38?

1 2 3 ▪ 109.82

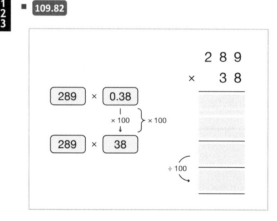

**14.** Calculate 2.69 x 1.84.

1 2 3 ▪ 4.9496

**15.** A tissue box is 12.02 centimetres (cm) in width. What is the total width of 13 tissue boxes? *Include the units cm in your answer.*

a b c ▪ 156.26 cm ▪ 156.26 centimetres

**16.** Nigel gets paid £8.66 per hour. He worked 35 hours last week. Calculate Nigel's wage for last week. *Include the £ sign in your answer.*

a b c ▪ £303.10

**17.** A grain of sand weighs 0.006 grams. What is the weight of 600,000 grains of sand? *Include the units g (grams) in your answer.*

a b c ▪ 3600 grams ▪ 3,600 g ▪ 3,600 grams ▪ 3600 g

**18.** Michael runs 10.6 miles each day. Michael ran 10.6 miles a day, 6 days per week for 8 weeks. How many miles did he run in total? *Don't include the units in your answer.*

1 2 3 ▪ 508.8

**19.** Alicia bought 4 cartons of soup for £1.99 each,
**a** and 7 tins of soup for £1.28 each. What was the
**b** total amount she spent?
**c** Include the £ sign in your answer.

▪ £16.92

**20.** Work out the answer to 2.13 x 4.7 x 1.8.
**a**
**b** ▪ 18.0198
**c**

**Level 3:**   Reasoning - Multiply integers and decimals
(column method).

❋ **Required:** 5/5   ❋ **Student Navigation:** on
❋ **Randomised:** off

**21.** The calculations below use the same
**a** digits but have different values.
**b** Can you explain a rule for positioning the decimal
**c** point?
16 x 25.4   = 406.4
1.6 x 2.54   = 4.064
0.16 x 2.54 = 0.4064

**22.** Imogen wants to know how much paint she needs
**a** to buy for four walls she is painting. The size of all
**b** the walls are the same: 10.2 metres (m) high by
**c** 2.6 m wide.
What is the total area of the walls in square
metres (m²).
*Don't include the units in your answer.*

▪ 106.08

**23.** Given that 93 x 24 = 2,232, what is 0.93 x 2.4?
**a**
**b** ▪ 2.232
**c**

**24.** Given that 93 x 24 = 2,232, what is 22.32 ÷ 24?
**a**
**b** ▪ 0.93
**c**

**25.** Benjamin says, '1.3 x 2.5 = 3.25'.
**a** Hajrah says, '1.3 x 2.5 = 32.5'.
**b** Who is correct? Explain your answer.
**c**

**Level 4:**   Problem Solving - Multiply integers and
decimals (column method).

❋ **Required:** 5/5   ❋ **Student Navigation:** on
❋ **Randomised:** off

**26.** A bottle of orange juice costs £1.55. If Danny
**a** drinks 3/4 of a bottle every week for a full year,
**b** how much will it cost him?
**c** *Include the £ sign in your answer.*

▪ £60.45

**27.** The rule for generating a number sequence is:
**a** *Multiply the previous number by 1.4 and then add*
**b** *6.*
**c** If the **2nd term** of the sequence is **17**, what is the
**difference** between the 3rd and the 4th terms?

▪ 17.92

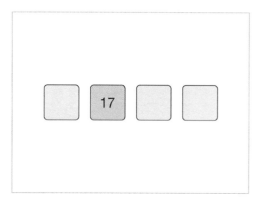

**28.** Mr Singh uses 967 units of electricity.
**a** He pays 26p per unit for the first 135 units and
**b** 19.5p per unit for the remaining units.
**c** How much is Mr Singh's electricity bill?
*Include the £ sign in your answer.*

▪ £197.34

**29.** Fill in the blanks in the column multiplication.
**a** What is the largest number you can make from the
**b** five missing numbers?
**c**

▪ 66440   ▪ 66,440

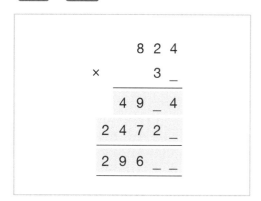

**30.** Calculate the area of the field not covered by the
**a** pond in square metres (m²).
**b** *Don't include the units in your answer.*
**c**

▪ 2518.762   ▪ 2,518.762

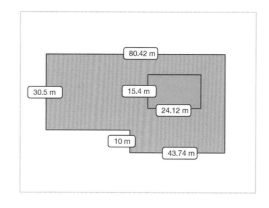

# Mathematics

**Y7**

## Fractions

Addition

Subtraction

Multiplication

Identify

Equivalent Fractions

Improper Fractions and Mixed Numbers

Decimals and Percentages

# Add and subtract fractions

**Competency:** Use the four operations, including formal written methods, applied to integers, decimals, proper and improper fractions and mixed numbers, all both positive and negative.

**Quick Search Ref:** 10009

Correct: Correct.   Wrong: Incorrect, try again?   Open: Thank you.

**Level 1:** Understanding - Terminology, adding and subtracting with like denominators and simplifying.

❋ **Required:** 7/10   ❋ **Student Navigation:** on   ❋ **Randomised:** off

---

**1.** What is the rule when adding or subtracting fractions?

☐
☒
☐
1/3

- Convert the fractions so that they have a common denominator and then add/subtract the denominators.
- Convert the fractions so that they have a common denominator and then add/subtract the numerators.
- Convert the fractions so that they have a common denominator and then add/subtract the numerators and add/subtract the denominators.

**2.** What is the **lowest common denominator** of the following fractions?
2/6, 3/8 and 5/12.

1
2
3

- 24

| multiples of | | |
|---|---|---|
| 6 | 8 | 12 |
| 6 | 8 | 12 |
| 12 | 16 | 24 |
| 18 | 24 | 36 |
| 24 | 32 | 48 |
| 30 | 40 | 60 |
| 36 | 48 | 72 |
| 42 | 56 | 84 |

**3.** 13/51 + 4/51 = 17/51.
What is the answer in its **simplest form**?

a
b
c

- 1/3

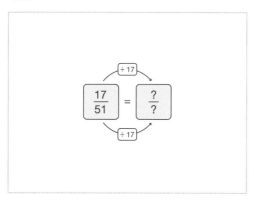

**4.** 4/16 + 2/16 + 3/16 = ___.

a
b
c

- 9/16

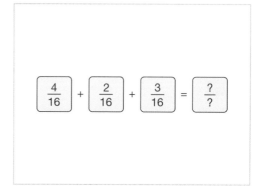

**5.** What is 3/12 - 1/12?
Give your answer in its **simplest form**.

a
b
c

- 1/6

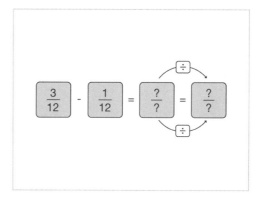

**6.** Calculate 2/5 + 1/3.

a
b
c

- 11/15

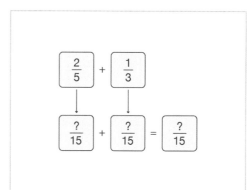

---

**Level 1:** *cont.*

**7.** What is 2/7 - 5/21?

 ■ 1/21

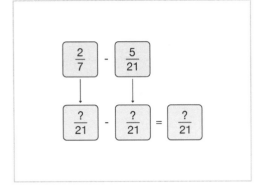

**8.** 17/20 + 1/20 = ___.
Give your answer in its **simplest form**.

 ■ 9/10

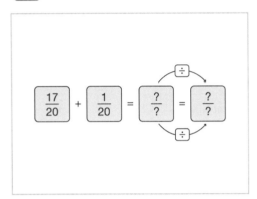

**9.** Calculate 5/9 - 2/9.
Give your answer in its **simplest form**.

■ 1/3

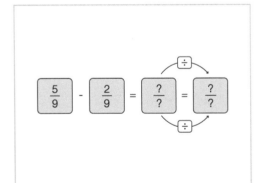

**10.** 7/12 + 3/8 = ___.

 ■ 23/24

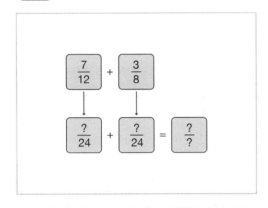

**Level 2:**   Fluency - Adding and subtracting with mixed number fractions and unlike denominators.

✵ **Required:** 7/10    ✵ **Student Navigation:** on
✵ **Randomised:** off

**11.** What is 4 3/10 + 3 4/10?
Give your answer as a **mixed number fraction**.

■ 7 7/10

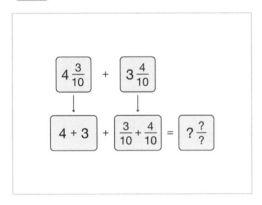

**12.** 3 4/7 - 1 1/7 = ___.
Give your answer as a **mixed number fraction**.

■ 2 3/7

**13.** 19/32 + 11/16 = ___.
Give your answer as a **mixed number fraction**.

■ 1 9/32

Level 2: *cont*.

**14.** Helen and Jess each have a cereal bar.
**a b c** Helen eats 13/15 and Jess eats 6/10.
How much do they eat altogether?
Give your answer as a mixed number fraction in its **simplest form**.

▪ **1 7/15**

**15.** In the following equation, what is the missing
**a b c** fraction?
13/7 - ___ = 18/14.
Give your answer in its **simplest form**.

▪ **4/7**

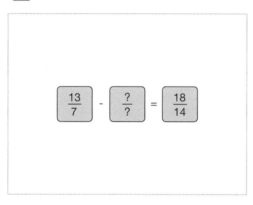

**16.** 13/18 - 4/27 ___.
**a b c** ▪ **31/54**

**17.** What is 5 1/12 + 1 6/18?
**a b c** Give your answer as a **mixed number fraction** in its **simplest form**.

▪ **6 5/12**

**18.** Ben lives 1 4/7 of a mile away from school.
**a b c** Joseph lives 3/5 of a mile away from school.
How **much further** does Ben live from school than Joseph?

▪ **34/35**

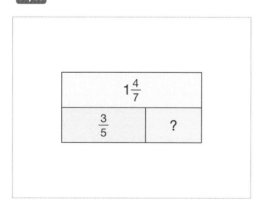

**19.** 5/7 - 2/3 = ___.
**a b c** ▪ **1/21**

**20.** 15/14 + 8/21 = ___.
**a** Give your answer as a **mixed number fraction**.
**b**
**c** ▪ **1 19/42**

**21.** Look at the image.
Select the **two fractions and a symbol** that make the statement true.

3/7  ▪ **-** ▪ + ▪ 8/9 ▪ **1/6** ▪ 2/3 ▪ 2/9 ▪ **13/18**

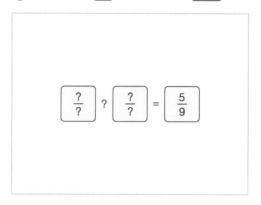

**22.** How **many parts** of the middle shape
**1** need shading to make the equation true?
**2**
**3** ▪ **8**

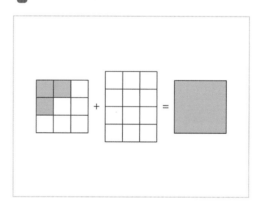

**23.** The diagram shows three fields.
Which has the **largest** perimeter?

▪ **Field (a)** ▪ Field (b) ▪ Field (c)

1/3

**Level 3:** *cont.*

**24.** Which letter has the **greatest** value?

■ x ■ y ■ z

1/3

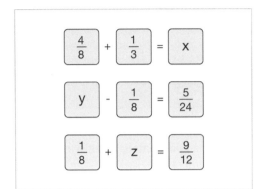

$$\frac{4}{8} + \frac{1}{3} = x$$

$$y - \frac{1}{8} = \frac{5}{24}$$

$$\frac{1}{8} + z = \frac{9}{12}$$

**25.** Miss Bloom conducted a survey of weekend bed
a   times. She says, "The **difference** between the
b   number of children who went to bed before 9:00
c   p.m. and those who went to bed after 9:00 p.m. is
    1/4".
    Is she correct? Explain how you know.

| time | fraction of children |
|---|---|
| before 8:00 p.m. | $\frac{3}{12}$ |
| 8:00 p.m. - 8:29 p.m. | $\frac{3}{24}$ |
| 8:30 p.m. - 8:59 p.m. | $\frac{1}{4}$ |
| 9:00 p.m. - 9:29 p.m. | $\frac{1}{6}$ |
| 9:30 p.m. - 10:00 p.m. | $\frac{1}{8}$ |
| after 10:00 p.m. | $\frac{1}{12}$ |

**Level 4:**   Problem Solving - Adding and subtracting
          fractions to solve problems.

❋ **Required:** 5/5   ❋ **Student Navigation:** on
❋ **Randomised:** off

**26.** Each fraction is the **sum** of the two fractions below
a   it.
b   What is the **sum of A, B and C?**
c

■ 33/84

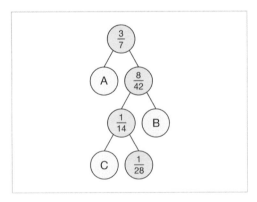

**27.** Find the difference in metres between the
a   perimeters of the 2 swimming pools.
b   Give your answer as a fraction.
c

■ 7/12

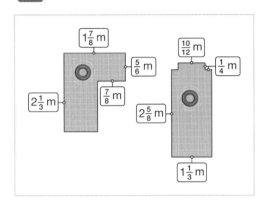

**28.** To open the safe, you need to press the buttons in
↑   the correct order. Use the clues to work out the
↓   correct order and then sort the button shapes first
    at the top, fifth at the bottom.
    **1st:** 1/6 + 7/18
    **2nd:** 1/4 + 7/24
    **3rd:** 5/6 - 1/4
    **4rd:** 3/4 - 5/36
    **5th:** 18/36 + 1/8

■ Triangle ■ Hexagon ■ Circle ■ Kite ■ Square

**29.** Add two side-by-side fractions to find the answer
a   to the box above.
b   What fraction goes in the box containing stars?
c   Give your answer as a mixed number fraction in its
    **simplest form.**

■ 2 1/3

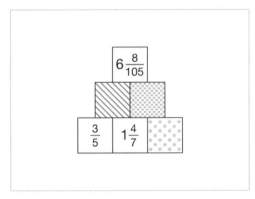

**30.** Calculate the fraction of the rectangle that is shaded.
Give your answer in its **simplest form**.

a
b
c

- 35/64

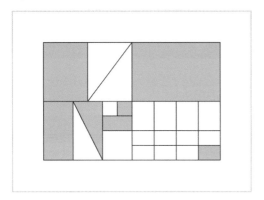

# Find a fraction of an amount

**Competency:**    Interpret fractions and percentages as operators.

**Quick Search Ref:**    10304

Correct: Correct.    Wrong: Incorrect, try again.    Open: Thank you.

**Level 1:**    Understanding - Methods and calculating fractions of simple amounts.

✹ **Required:** 7/10    ✹ **Student Navigation:** on    ✹ **Randomised:** off

**1.**    What is 1/3 of 15?

[1 2 3]    ▪ **5**

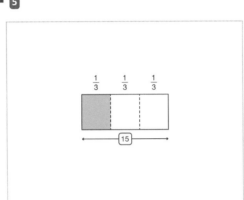

**2.**    What is 1/4 of 9?

[a b c]    ▪ **2 1/4**   ▪ **2.25**   ▪ **9/4**

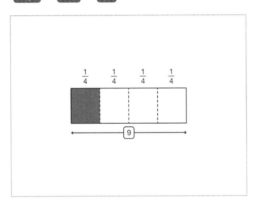

**3.**    Select the **two** methods that can be used to find a fraction of an amount.

2/4

- Divide the amount by the numerator and denominator.
- **Multiply the amount by the numerator and then divide by the denominator.**
- Multiply the amount by the numerator and denominator.
- **Divide the amount by the denominator and then multiply by the numerator.**

**4.**    To calculate 2/5 of 7, which method is most appropriate?

1/2

- **Multiply the amount by the numerator and then divide by the denominator.**
- Divide the amount by the denominator and then multiply by the numerator.

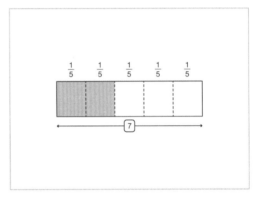

**5.**    To calculate 2/3 of 18, which method is most appropriate?

1/2

- **Divide the amount by the denominator and then multiply by the numerator.**
- Multiply the amount by the numerator and then divide by the denominator.

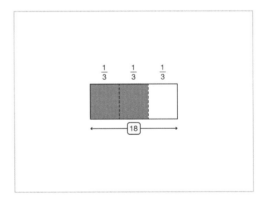

**6.** What is 4/5 of 20?

 ▪ 16

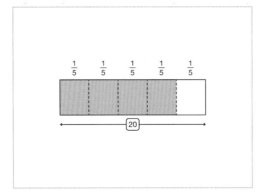

**7.** Calculate 3/4 of £24.

 *Include the £ sign in your answer.*

▪ £18.00  ▪ £18

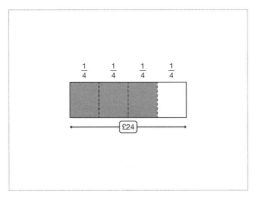

**8.** To calculate 5/6 of 13, which method is most appropriate?

1/2

▪ Multiply the amount by the numerator and then divide by the denominator.

▪ Divide the amount by the denominator and then multiply by the numerator.

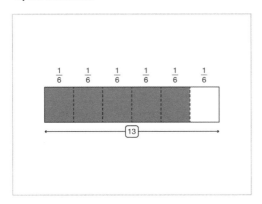

**9.** What is 1/2 of 17?

 ▪ 8 1/2  ▪ 8.5

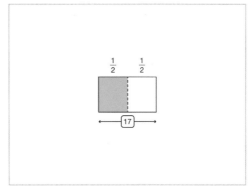

**10.** Calculate 2/5 of 35.

 ▪ 14

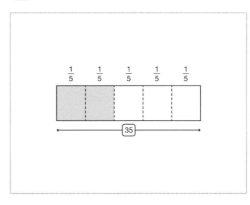

**Level 2:**  Fluency - Calculating fractions of amounts in context and with units (fractional and decimal answers).

✱ **Required:** 7/10   ✱ **Student Navigation:** on
✱ **Randomised:** off

**11.** James has 40 millilitres of water in a test tube but he spills 3/8 of it. How much does he spill?
*Include the unit ml (millilitres) in your answer.*

▪ 15 ml  ▪ 15 millilitres

**12.** Calculate 1/5 of £2.
*Include the unit p (pence) in your answer.*

▪ 40p  ▪ 40 pence

**13.** Calculate 5/8 of 18.
Give your answer as a mixed number fraction in its simplest form.

▪ 11 1/4

**14.** What is 3/5 of 13 as a decimal?

▪ 7.8

Level 2: *cont*.

**15.** What is the missing **fraction** in the following calculation?

_ of 70 = 14

Give your answer in its simplest form.

■ 1/5

**16.** Kuran gives away 3/7 of his apples. How many does he have left?

■ 12

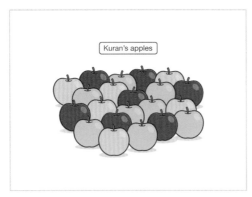

Kuran's apples

**17.** What is 5/6 of 15 as a decimal?

■ 12.5

**18.** A box contains 28 pencils. Mr Jones gives out 2/7 to his pupils. How many are left in the box?

■ 20

**19.** Calculate 1/8 of 1 kg.

*Include the units g (grams) in your answer.*

■ 125 g ■ 125 grams

**20.** Calculate 3/4 of 22.

Give your answer as a mixed number fraction in its simplest form.

■ 16 1/2

**Level 3:** Reasoning - Comparing, reasoning and inverse calculations.

✿ **Required:** 5/5   ✿ **Student Navigation:** on
✿ **Randomised:** off

**21.** Amy says, "To find 3/7 of 28, I find 1/7 and then multiply my answer by 3".

Is she correct? Explain your answer.

**22.** Calculate 3/5 of 37.5.

■ 22.5

**23.** Which symbol makes the statement true?

5/6 of 42 _ 6/7 of 42.

■ < ■ = ■ >

1/3

**24.** 30 is 5/8 of what an amount?

■ 48

**25.** 4/9 of Esmee's marbles are shown in the image. How many marbles does she have in total?

■ 54

**Level 4:** Problem Solving - With fractions of amounts.

✿ **Required:** 5/5   ✿ **Student Navigation:** on
✿ **Randomised:** off

**26.** Katy is making tzatziki. How much cucumber will she need?

*Include the units g (grams) in your answer.*

■ 62.5 grams   ■ 62.5 g

| 300 g recipe contains: | |
|---|---|
| $\frac{1}{3}$ | Greek yogurt |
| $\frac{5}{24}$ | cucumber |
| $\frac{1}{8}$ | lemon juice |
| $\frac{2}{15}$ | garlic |
| $\frac{1}{6}$ | dill |
| $\frac{1}{30}$ | salt |

**27.** Use the digits **2**, **3**, **4** and **7** to make the equation true. What does each side of the equation equal?

▪ 28  ▪ 21

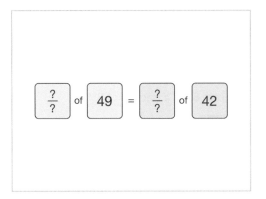

$$\frac{?}{?} \text{ of } 49 = \frac{?}{?} \text{ of } 42$$

**28.** Move from **start** to **finish** and arrange the patterns to represent the path you take.

• You can only pass through calculations that have an integer as the answer.

• You can only move left/right/up/down.

▪ Stripes ▪ Dots ▪ Zigzagss ▪ Checks ▪ Waves

| start | $\frac{1}{3}$ of 12 | $\frac{3}{8}$ of 74 | $\frac{5}{9}$ of 35 |
|---|---|---|---|
| $\frac{1}{4}$ of 13 | $\frac{1}{4}$ of 72 | $\frac{6}{8}$ of 84 | $\frac{1}{6}$ of 38 |
| $\frac{2}{9}$ of 20 | $\frac{5}{6}$ of 80 | $\frac{4}{9}$ of 54 | $\frac{5}{7}$ of 90 |
| $\frac{7}{8}$ of 44 | $\frac{3}{5}$ of 24 | $\frac{5}{12}$ of 48 | finish |

**29.** Ryan is having a dinner party. 861 grams of rice is enough to serve his guests.

If the number of guests increases by 2/3, how much rice is needed?

*Include the units kg (kilograms) in your answer.*

▪ 1.435 kg  ▪ 1.435 kilograms

**30.** The children in Year 7 have to join a sports team.

1/3 choose football.

1/6 choose badminton.

3/8 choose gymnastics.

The remaining 27 children choose karate.

How many children are in Year 7?

▪ 216

# Express One Quantity as a Fraction of Another

**Competency:**    Express one quantity as a fraction of another where the fractions are less than 1 and greater than 1.

**Quick Search Ref:**    10267

Correct: Correct.    Wrong: Incorrect, try again.    Open: Thank you.

**Level 1:**    Understanding - Understanding and expressing quantities as a fraction of another.

✿ Required: 7/10    ✿ Student Navigation: on    ✿ Randomised: off

**1.** What does the fraction **7/10** represent?

 ■ 7 out of 10 ■ 10 out of 7 ■ 7 out of 17

1/3

$$\frac{7}{10}$$

**2.** What is 13 out of 30 as a fraction?

 ■ 13/30

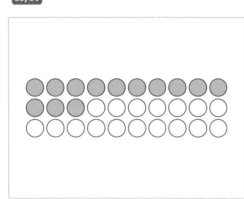

**3.** What fraction of the grid is shaded?

 ■ 11/20

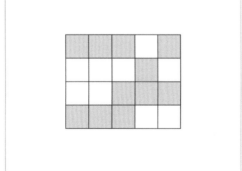

**4.** Express 62 as a fraction of 85.

 ■ 62/85

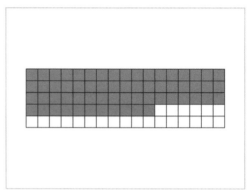

**5.** Give 63 as a fraction of 50.

 ■ 63/50

**6.** Express 46 out of 60 as a fraction.
*Give your answer in its simplest form.*

 ■ 23/30

**7.** What is 78 out of 22 as a **mixed number fraction** in its simplest form?

 ■ 39/11 ■ 3 6/11 ■ 78/22 ■ 3 12/22

1/4

**8.** What does the fraction 15/19 represent?

 ■ 19 out of 15 ■ 15 out of 34 ■ 15 out of 19

1/3

**9.** What is 15 out of 90 as a fraction?
*Give your answer in its simplest form.*

 ■ 1/6

**Level 1:** *cont.*

**10.** What fraction of the grid is shaded?

  ■ 6/15

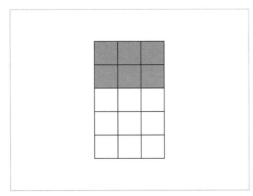

**Level 2:** Fluency - Expressing quantities as a fraction of another quantity in context.

✹ **Required:** 7/10    ✹ **Student Navigation:** on
✹ **Randomised:** off

**11.** What is £80 out of £640 as a fraction?
*Give your answer in its simplest form.*

abc  ■ 1/8

**12.** Lennie makes 390 bracelets and sells 240 at a village fête. What **fraction** of the bracelets does Lennie **sell**?
*Give your answer in its simplest form.*

■ 8/13

**13.** There are 28 children in class 7 and 14 have school lunches.
What fraction of the children have school lunches?
*Give your answer in its simplest form.*

■ 1/2

**14.** There are 15 girls and 18 boys in a drama club.
What fraction of the drama club is girls?

abc  ■ 15/33  ■ 5/11

**15.** Liz buys a car for £1,200 and then sells it for £1,800.
What is the sale price as a fraction of the original price?
*Give your answer as a mixed number fraction in its simplest form.*

■ 1 1/2

**16.** Express 120 grams as a fraction of 2.6 kilograms.
*Give your answer in its simplest form.*

abc  ■ 3/65

**17.** A pack of grapes contains 58 green grapes and 72 red grapes. What fraction of the grapes are red?
*Give your answer in its simplest form.*

■ 36/65

**18.** What is 85p as a fraction of £4.00?
*Give your answer in its simplest form.*

abc  ■ 17/80

**19.** What is 300 as a fraction of 120?
*Give your answer as a mixed number fraction in its simplest form.*

abc  ■ 2 1/2

**20.** Danny has 18 marbles and gives 7 to Mark. What fraction of the marbles does Mark receive?

abc  ■ 7/18

**Level 3:** Reasoning - Ordering, comparing and explaining quantities as fractions of other quantities.

✹ **Required:** 5/5    ✹ **Student Navigation:** on
✹ **Randomised:** off

**21.** Select the symbol that makes the statement true.
56 out of 60 __ 78 out of 90.

■ < ■ = ■ >

1/3

**22.** Three friends played a computer game.
Aliya scored 18 out of 20.
Tania scored 26 out of 30
Scott scored 35 out of 40.
1/3    Who got the largest fraction of points?

■ Aliya ■ Tania ■ Scott

**23.** Hector flipped a coin lots of times.
The ratio of heads to tails was 42:37.
Hector says the fraction of coin flips landing on tails was 37/42.
What mistake has he made?

**24.** Max, Tom and Katie are making lemonade using lemon juice and water.
Whose lemonade is stronger?
Max: 24 litres of lemon juice, 21 litres of water.
1/3    Tom: 36 litres of lemon juice, 24 litres of water.
Kate: 49 litres of lemon juice, 41 litres of water.

■ Max ■ Tom ■ Kate

**25.** Harvey has been at work for 90 minutes, which represents 1/5 of his working day. How many **hours** does Harvey spend at work **each day**?
*Give your answer as a decimal and don't include the units.*

■ 7.5

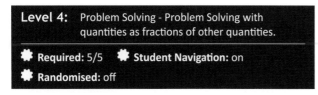

**Level 4:** Problem Solving - Problem Solving with quantities as fractions of other quantities.

❋ **Required:** 5/5   ❋ **Student Navigation:** on
❋ **Randomised:** off

**26.** Bobby is rolling a 6-sided die and he records the number of times he rolls each number.
**a b c** What fraction of the rolls land on a **prime number**?
1: 12 rolls.
2: 13 rolls.
3: 8 rolls.
4: 15 rolls.
5: 7 rolls.
6: 9 rolls.
*Give your answer in its simplest form.*

▪ 7/16

**27.** Arrange Roxy's test scores in order, starting with her worst score (the lowest fractional score).

↑ ↓   ▪ Biology ▪ French ▪ Maths ▪ English ▪ History

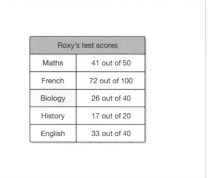

| Roxy's test scores | |
|---|---|
| Maths | 41 out of 50 |
| French | 72 out of 100 |
| Biology | 26 out of 40 |
| History | 17 out of 20 |
| English | 33 out of 40 |

**28.** Move from **start** to **finish** and arrange the patterns to represent the path you take.
↑ ↓   • You can only pass through fractions that simplify.
• You can only move left/right/up/down.

▪ Checked ▪ Zigzag ▪ Dotted ▪ Striped ▪ Waves

| start | 55 as a fraction of 94 | $\frac{10}{50}$ | 14p out of 37p |
|---|---|---|---|
| 72 out of 90 | $\frac{628}{317}$ | 31 as a fraction of 27 | $\frac{98}{157}$ |
| 64 cm out of 184 cm | 112 as a fraction of 280 | $\frac{95}{270}$ | 85 as a fraction of 174 |
| $\frac{113}{117}$ | 57 out of 64 | 16 ml out of 128 ml | finish |

**29.** There are 570 pupils at a school.
**a b c** 120 pupils are in Year 7.
2/5 of Year 7 pupils are boys.
What fraction of the **whole school** is represented by Year 7 girls?
*Give your answer in its simplest form.*

▪ 12/95

**30.** There are 120 guests at a wedding.
**a b c** 52 of the guests are male.
26 of the guests are vegetarian.
41 of the male guests are not vegetarian.
What fraction of the guests is represented by female vegetarians?
*Give your answer in its simplest form.*

▪ 1/8

|  | vegetarian | non-vegetarian | total |
|---|---|---|---|
| male | | | |
| female | | | |
| total | | | |

# Compare and order fractions using symbols

**Competency:** Order and compare positive and negative integers using the symbols =, ≠, <, >, ≤, ≥.

**Quick Search Ref:** 10082

Correct: Correct.    Wrong: Incorrect, try again.    Open: Thank you.

**Level 1:** Understanding - Terminology and comparing fractions of shapes.

✱ **Required:** 7/10    ✱ **Student Navigation:** on    ✱ **Randomised:** off

---

**1.** What is a **common multiple**?

1/4

- A fraction that is equal to or larger than one whole.
- A number which is a factor of two or more numbers.
- **A number that is a multiple of two or more numbers.**
- A fraction with the same value as another. For example, one-half and two-quarters.

**2.** Which **three** fractions have denominators with a common multiple less than 20?

3/6

- **2/3** ▪ 5/14 ▪ 1/7 ▪ **1/6** ▪ **2/12** ▪ 3/5

| multiples (less than 20) | | | | | |
|---|---|---|---|---|---|
| 3 | 14 | 7 | 6 | 12 | 5 |
| 3 | 14 | 7 | 6 | 12 | 5 |
| 6 | | 14 | 12 | | 10 |
| 9 | | | 18 | | 15 |
| 12 | | | | | |
| 15 | | | | | |
| 18 | | | | | |

**3.** Which is the **largest** fraction: 3/9 or 3/7?

1/2

- **3/7** ▪ 3/9

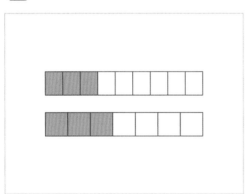

**4.** Which is the **smallest** fraction: 5/8 or 5/3?

1/2

- **5/8** ▪ 5/3

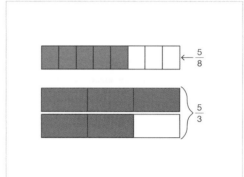

$\leftarrow \frac{5}{8}$

$\frac{5}{3}$

**5.** To compare the following fractions they need to have the same denominator. What is the denominator?
3/4, 5/8, 1/3, 11/12

- **24**

| multiples of | | | |
|---|---|---|---|
| 4 | 8 | 3 | 12 |
| 4 | 8 | 3 | 12 |
| 8 | 16 | 6 | 24 |
| 12 | 24 | 9 | 36 |
| 16 | 32 | 12 | 48 |
| 20 | 40 | 15 | 60 |
| 24 | 48 | 18 | 72 |
| 28 | 56 | 21 | 84 |
| 32 | 64 | 24 | 96 |

**6.** How many parts of shape (b) do you need to shade to make a fraction equivalent to shape (a)?

- **9**

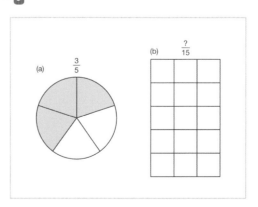

(a) $\frac{3}{5}$    (b) $\frac{?}{15}$

**7.** Sort the symbols in the following order:

Less than or equal to
Equal to
Greater than
Not equal to
Less than
Greater than or equal to

■ ≤ ■ = ■ > ■ ≠ ■ < ■ ≥

**8.** Which is the largest fraction?

■ 3/4 ■ 5/8

1/2

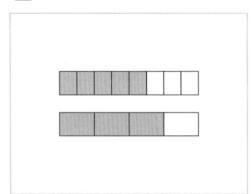

**9.** Select the **two** fractions that are greater than 6/12.

■ 3/24 ■ 8/10 ■ 5/18 ■ 5/8 ■ 1/2

2/5

**10.** Which is the smallest fraction?

■ 3/8 ■ 1/2

1/2

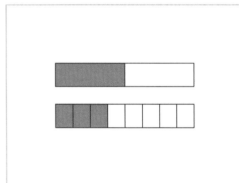

**11.** Which fraction is greatest?
7/18, 16/36 or 3/6

■ 3/6 ■ 7/18 ■ 16/36

1/3

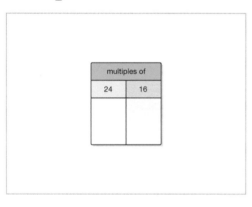

| multiples of | | |
|---|---|---|
| 18 | 36 | 6 |
| | | |

**12.** Select the symbol that makes the following statement true:
17/24 __ 11/16

1/3   ■ < ■ = ■ >

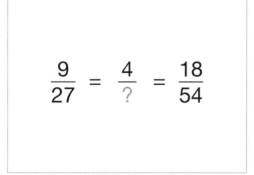

| multiples of | |
|---|---|
| 24 | 16 |
| | |

**13.** Select the fraction that is ≠ 28/32?

■ 7/8 ■ 14/16 ■ 3/4 ■ 21/24

1/4

**14.** What missing number makes the following statement true?
9/27 = 4/__ = 18/54

■ 12

$$\frac{9}{27} = \frac{4}{?} = \frac{18}{54}$$

**Level 2:** *cont.*

**15.** Arrange the following fractions in ascending order (smallest first).

↑
↓
▪ 5/32 ▪ 1/2 ▪ 9/16 ▪ 5/8 ▪ 3/4 ▪ 7/8 ▪ 2/2

**16.** Select the **three** fractions that are ≤ 15/21.

▪ 7/7 ▪ 10/14 ▪ 2/7 ▪ 18/21 ▪ 12/21

3/5

**17.** Select the **three** fractions that have a value > 26/78.

▪ 1/3 ▪ 8/9 ▪ 4/3 ▪ 2/6 ▪ 1/2

3/5

**18.** Select the symbol that makes the following statement true.
17/34 ___ 27/54

1/3  ▪ < ▪ = ▪ >

**19.** Arrange the fractions in descending order (largest first).

↑
↓
▪ 15/30 ▪ 2/5 ▪ 4/15 ▪ 1/10

**20.** Which fraction is the **smallest**: 3/5, 7/10 or 19/25?

▪ 3/5 ▪ 7/10 ▪ 19/25

1/3

| multiples of | | |
|---|---|---|
| 5 | 10 | 25 |
| | | |

**Level 3:** Reasoning - Convert, order and compare fractions.

✹ **Required:** 5/5   ✹ **Student Navigation:** on
✹ **Randomised:** off

**21.** Which statement is **incorrect**?

▪ 2/17 < 6/34 ▪ 4/9 = 24/54 ▪ 2/3 ≠ 168/252
▪ 19/20 > 91/100

1/4

**22.** Dave says, "10 is greater than 9, therefore 1/10 is greater than 1/9".
Is he correct? Explain your answer.

a
b
c

**23.** Which **three** fractions are equal to the fraction shown by the arrow?

3/6

▪ 3/4 ▪ 6/16 ▪ 7/10 ▪ 15/40 ▪ 27/72 ▪ 8/16

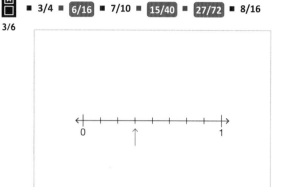

**24.** Bailey says, "I can compare these fractions by simplifying instead of converting them to fractions with a common denominator".
Is he correct? Explain your answer.

a
b
c

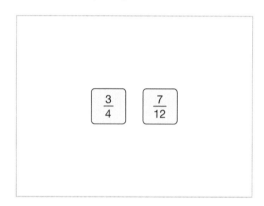

$$\frac{3}{4} \qquad \frac{7}{12}$$

**25.** Using each of the digit cards **once**, what is the largest **proper fraction** you can make?

a
b
c

▪ 73/81

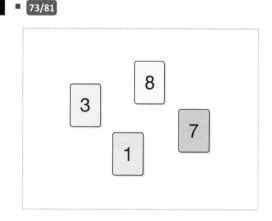

**Level 4:** Problem Solving - Fractions with denominators of shared multiples.

❋ **Required:** 5/5 ❋ **Student Navigation:** on
❋ **Randomised:** off

**26.** Which **two** fractions make one whole when added together?

■ 6/24 ■ 1/18 ■ 1/3 ■ 15/18 ■ 3/9 ■ 4/8 ■ 12/72

2/7

| $\frac{4}{8}$ | $\frac{12}{72}$ | $\frac{3}{9}$ |
|---|---|---|
| $\frac{1}{72}$ | $\frac{15}{18}$ | $\frac{1}{18}$ |
| $\frac{1}{3}$ | $\frac{2}{9}$ | $\frac{6}{24}$ |

**27.** The fractions are shown in ascending order (smallest first). What is the missing numerator?

■ 5

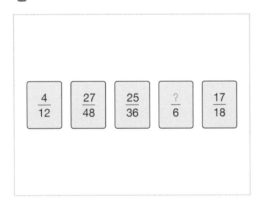

| $\frac{4}{12}$ | $\frac{27}{48}$ | $\frac{25}{36}$ | $\frac{?}{6}$ | $\frac{17}{18}$ |

**28.** For homework Alex reads:
96 pages of her 384 page French book.
112 out of 280 pages in her History book.
104 out of 182 pages of her Chemistry book.

1/3  Which book did she read the greatest fraction of?

■ French ■ History ■ Chemistry

**29.** Katie shares a bag of nuts with her friends. She keeps 9/34 and gives 3/17 to Kai, 8/34 to George, 12/102 to Freddie and 14/68 to Becca.
*Arrange the names in ascending order (starting with the person who has the least nuts).*

■ Freddie ■ Kai ■ Becca ■ George ■ Katie

**30.** Darcey and Levi live on the same road as the church.
a
b
c  Darcey's house is 24/58 of a mile away from the church and 75/87 away from Levi's house.
How far away is Levi's house from the church?

■ 13/29

# Identify and use equivalent fractions

**Competency:**     Find equivalent fractions, simplify fractions and solve problems involving fractions.

**Quick Search Ref:**     10055

Correct: Correct     Wrong: Incorrect, try again.          Open: Thank you.

**Level 1:**     Understanding - Terminology, equivalence and simplification.

✿ **Required:** 7/10          ✿ **Student Navigation:** on          ✿ **Randomised:** off

---

**1.** What are equivalent fractions?

1/4

- The bottom number in a fraction, which shows how many equal parts the whole is divided into.
- **Fractions that have the same value but are shown with different numbers.**
- The top number in a fraction, which shows how many parts of the whole there are.
- Fractions which have been reduced to the smallest possible whole number.

$$\frac{1}{2} = \frac{2}{4}$$

**2.** How do you fully simplify a fraction?

1/3

- Divide the numerator and denominator by the same number until the numerator becomes 1.
- **Divide the numerator and denominator by the highest common factor.**
- Divide the denominator by the numerator.

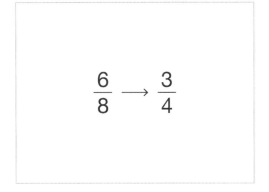

$$\frac{6}{8} \longrightarrow \frac{3}{4}$$

**3.** What is the missing numerator?

- **3**

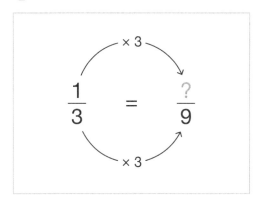

**4.** Taylor says, "To compare fractions, I can simplify them instead of finding a common denominator". Is this true?

1/3
- **Sometimes** ▪ Always ▪ Never

**5.** Choose the **three** fractions that are equivalent to 5/15.

3/7
- ▪ **20/60** ▪ 15/35 ▪ **30/90** ▪ 10/15 ▪ 60/150
- ▪ **120/360** ▪ 25/60

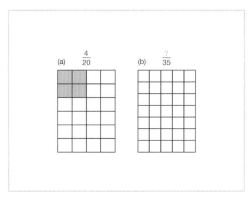

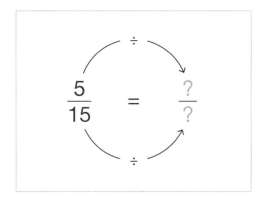

---

**6.** Select the **three** numbers that are divisible by 3.

3/5

▪ 54 ▪ 132 ▪ 26 ▪ 49 ▪ 18

**7.** What is 18/21 in its simplest form?

a b c ▪ 6/7

| factors of | |
|---|---|
| 18 | 21 |
| 1 | 1 |
| 2 | 3 |
| 3 | 7 |
| 6 | |
| 9 | |

**8.** Which fraction is equivalent to 10/15?

1/3

▪ 2/3 ▪ 5/8 ▪ 4/5

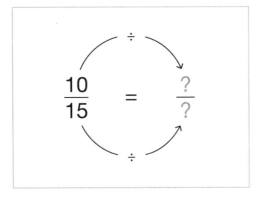

**9.** Select the **three** fractions that are equivalent to 2/7.

3/7

▪ 20/70 ▪ 4/28 ▪ 5/35 ▪ 4/12 ▪ 46/161 ▪ 8/40
▪ 6/21

**10.** What is 33/99 in its simplest form?

a b c ▪ 1/3

---

**Level 2:** Fluency - Recognising and finding equivalent fractions in numbers and shapes.

❋ **Required:** 7/10    ❋ **Student Navigation:** on
❋ **Randomised:** off

**11.** How many parts of the grid do you need to shade to make a fraction equivalent to 6/8?

1 2 3 ▪ 27

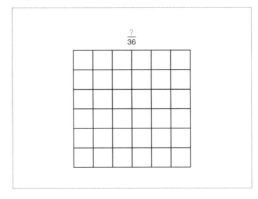

**12.** Select the three fractions that are equal to 5/75.

3/7

▪ 7/105 ▪ 3/50 ▪ 10/150 ▪ 9/27 ▪ 1/30 ▪ 2/15
▪ 60/900

$$\frac{5}{75} = \frac{?}{?}$$

**13.** What is 153/171 in its simplest form?

a b c ▪ 17/19

**14.** What missing numerator makes the following statement true?
34/119 = ?/7

1 2 3

▪ 2

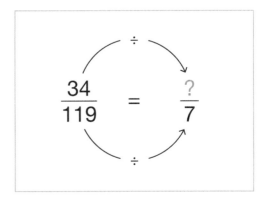

---

**Level 2:** *cont.*

**15.** Which **two** fractions are equivalent to the fraction indicated by the arrow?

 **21/56** ▪ **5/16** ▪ **51/136** ▪ **9/27** ▪ **16/63**

2/5

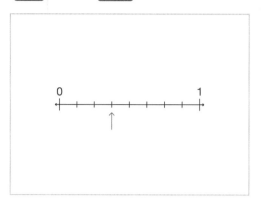

**16.** What is the missing denominator?

462/? = 154/171

▪ **513**

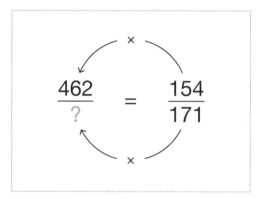

**17.** How many parts of the circle do you need to shade to show a fraction equivalent to 21/56?

 ▪ **6**

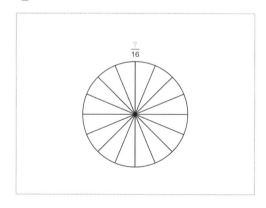

**18.** Three-quarters is equal to how many seventy-sixths?

*Give your answer in digits.*

▪ **57**

$$\frac{3}{4} = \frac{?}{76}$$

**19.** What missing denominator makes the following statement true?

1/? = 9/54

▪ **6**

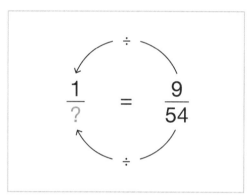

**20.** What is 4/72 in its simplest form?

 ▪ **1/18**

**Level 3:**   Reasoning - Reasoning with equivalent fractions.

✸ **Required:** 5/5    ✸ **Student Navigation:** on
✸ **Randomised:** off

**21. Using every digit card once**, make a fraction equivalent to 72/108.

 ▪ **18/27**

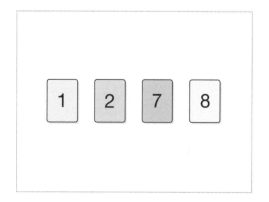

**Level 3: cont.**

**22.** Sonny simplifies 16/48 by dividing by 2 each time. Ben says he can simplify 16/48 in one step. Is Ben correct? Explain how you know.

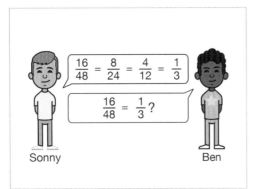

Sonny                    Ben

**23.** Which symbol makes the following statement true?

■ < ■ = ■ >

1/3

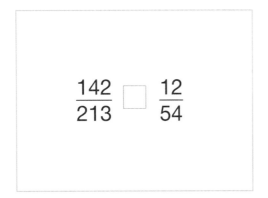

$$\frac{142}{213} \square \frac{12}{54}$$

**24.** Which fraction is the odd one out?

■ 16/24 ■ 7/36 ■ 12/168 ■ 15/75 ■ 21/315

1/5

**25.** Amalia says that 13/104 is impossible to simplify because 13 is a prime number. Is she correct? Explain your answer.

**Level 4:** Problem solving - Problems involving equivalent fractions and simplifying.

✹ **Required:** 5/5    ✹ **Student Navigation:** on
✹ **Randomised:** off

**26.** Stan and Jasper both scored 60% on their tests. Stan's test was marked out of 30. Jasper's test was marked out of 40. How many more marks did Jasper get than Stan?

■ 6

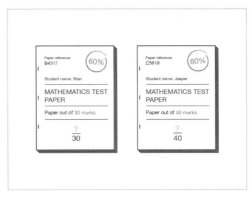

**27.** Calculate each numbered part as a fraction of the shape. Which **three** parts are equivalent fractions?

■ 1 ■ 2 ■ 3 ■ 4 ■ 5 ■ 6 ■ 7

3/7

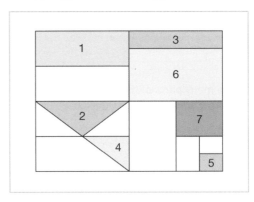

**28.** Emma gives Adam 16/40 of her counters. She gives Brooke 14/85. What fraction must she give to Jon to make the **total** of Jon's and Brooke's counters **equivalent** to Adam's?
*Give your answer in its simplest form.*

■ 4/17

**29.** Ryder is thinking of a fraction that has a denominator **less than 30**.

His fraction is equivalent to 82/205.
How many possible fractions could he be thinking of?

- [5]

$$\frac{?}{<30} = \frac{82}{205}$$

**30.** Norah makes a batch of lemonade which is made up of 3 parts lemon to 2 parts water.

Craig makes a batch of lemonade which is 7 parts lemon to 6 parts water.

**How much more** lemon is in Norah's lemonade than in Craig's?

*Give your answer as a fraction in its simplest form.*

- [4/65]

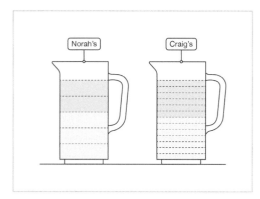

# Simplify fractions and express fractions with the same denominator

**Competency:** Use common factors to simplify fractions; use common multiples to express fractions in the same denomination.

**Quick Search Ref:** 10075

Correct: Correct.    Wrong: Incorrect, try again.        Open: Thank you.

**Level 1:**   Understanding - Identifying the highest common factor to simplify fractions.

✼ **Required:** 7/10        ✼ **Student Navigation:** on        ✼ **Randomised:** off

**1.** **Equivalent fractions** are:

■ fractions equal to or larger than one whole.
■ fractions that are less than one whole.
1/4   ■ fractions which have the same value but are shown with different numbers.
■ a whole number and a proper fraction combined.

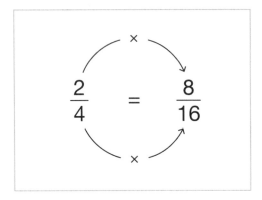

**2.** The **highest common factor** is:

■ a whole number that divides exactly into another whole number without a remainder.
1/4   ■ the largest number that divides exactly into two or more numbers.
■ the smallest positive number that is a multiple of two or more numbers.
■ two or more fractions which have the same value. For example, 1/2 = 2/4.

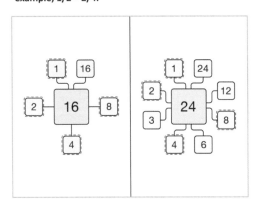

**3.** How do you **fully simplify** a fraction?

■ Multiply the numerator and denominator by the same number.
1/3   ■ Divide the numerator and denominator by the highest common factor.
■ Divide the denominator by the numerator.

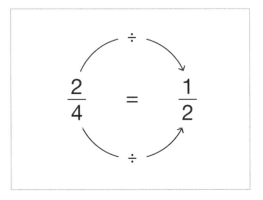

**4.** What is the highest common factor of 16 and 24?

■ 2  ■ 4  ■ 8  ■ 16  ■ 24

1/5

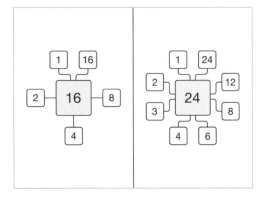

**5.** Simplify the fraction 6/10.

■ 3/5

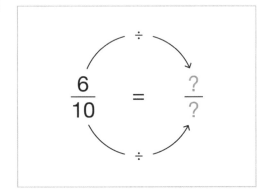

**6.** To convert two or more fractions so that they have a common denominator, you need to:

1/4

- find the highest common factor.
- **find the lowest common multiple of the denominators.**
- multiply the denominators.
- write each fraction in its simplest form.

**7.** What is the **lowest common multiple** of the three denominators shown?

1/4

- 4 ■ **8** ■ 16 ■ 32

$$\frac{1}{2} \qquad \frac{3}{4} \qquad \frac{3}{8}$$

**8.** To simplify 12/18, what number should you divide the numerator and denominator by?

1
2
3

■ **6**

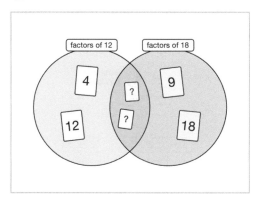

**9.** Write 4/20 in its simplest form.

a
b
c

■ **1/5**

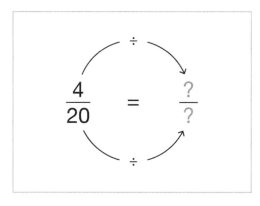

**10.** Simplify the fraction 9/21.

a
b
c

■ **3/7**

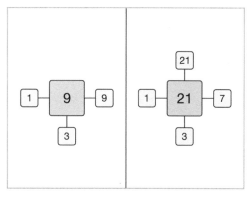

**Level 2:** Fluency - Simplifying fractions and comparing fractions with different denominators.

✱ **Required:** 7/10     ✱ **Student Navigation:** on
✱ **Randomised:** off

**11.** What is 28/36 in its simplest form?

a
b
c

■ **7/9**

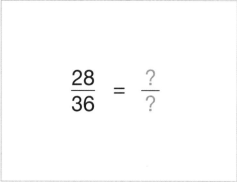

$$\frac{28}{36} = \frac{?}{?}$$

**12.** Sort the fractions in order of value starting with the lowest first.

↑
↓

■ **3/8** ■ **20/32** ■ **11/16** ■ **3/4** ■ **7/8**

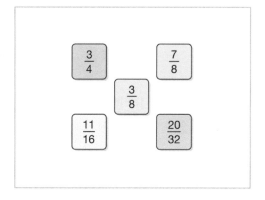

**Level 2:** *cont.*

**13.** What is 27/45 in its simplest form?

 ■ 3/5

$$\frac{27}{45} = \frac{?}{?}$$

**14.** Which fraction is **not** written in its simplest form?

■ 7/9 ■ 4/11 ■ 9/12 ■ 13/15

1/4

$$\frac{7}{9} \qquad \frac{4}{11}$$

$$\frac{9}{12} \qquad \frac{13}{15}$$

**15.** What is the missing denominator?

 ■ 27

$$\frac{2}{3} = \frac{18}{?}$$

**16.** Which fraction has the **highest** value?

■ 2/5 ■ 8/10 ■ 3/15 ■ 12/20

1/4

$$\frac{2}{5} \qquad \frac{8}{10}$$

$$\frac{3}{15} \qquad \frac{12}{20}$$

**17.** Which symbol makes the statement true?

 ■ > ■ < ■ =

1/3

$$\frac{2}{5} \quad \square \quad \frac{3}{7}$$

**18.** Write 35/70 in its simplest form.

■ 1/2

$$\frac{35}{70} = \frac{?}{?}$$

**19.** Which fraction is **not** written in its simplest form?

■ 4/5 ■ 7/10 ■ 11/15 ■ 15/20

1/4

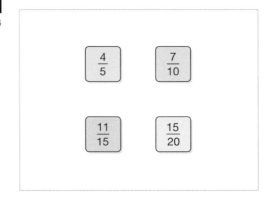

$$\frac{4}{5} \qquad \frac{7}{10}$$

$$\frac{11}{15} \qquad \frac{15}{20}$$

**20.** Sort the fractions in order of value (smallest first).

■ 7/24 ■ 5/12 ■ 2/3 ■ 7/8

$$\frac{2}{3} \qquad \frac{7}{8}$$

$$\frac{5}{12} \qquad \frac{7}{24}$$

**Level 3:** Reasoning - Comparing and simplifying fractions within a question.

✱ **Required:** 5/5    ✱ **Student Navigation:** on
✱ **Randomised:** off

**Level 4:** Problem Solving - Simplifying fractions to solve problems.

✱ **Required:** 5/5    ✱ **Student Navigation:** on
✱ **Randomised:** off

**21.** There are 72 children in year 6 and 48 are boys. What fraction of the children are **girls**?
*Give your answer in its simplest form.*

a b c

▪ 1/3

**22.** Julie and Syed are working out the answer to the question 27/54 + 41/82.
Julie says, "To add the fractions you need to convert both of the fractions so they have a common denominator".
Syed says, "There's an easier way to calculate the answer to this question".
What is Syed's method?

a b c

**23.** There are 5 pairs of fractions shown on 11 fraction cards.
Each pair are **equivalent fractions** and one of the fractions in the pair is written in its **simplest form**.
What is the value of the fraction card **not** in a pair?

a b c

▪ 8/11

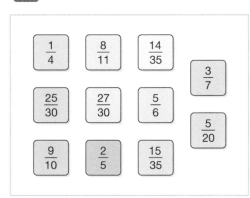

**24.** Which shape has **3/8** of its area shaded?

1/4

▪ (a) ▪ (b) ▪ (c) ▪ **(d)**

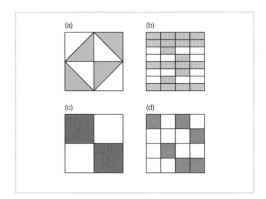

**25.** Eric says, "If the numerator and denominator of a fraction are even, you can divide them both by 2 repeatedly until the fraction is in its simplest form".
Is Eric correct? Explain your answer.

a b c

**26.** Phoebe is thinking of a fraction that is equivalent to 66/102. The denominator is less than 20. What fraction is Phoebe thinking of?

a b c

▪ 11/17

**27.** 52/65 = 4/x = y/z.
x + z = 50.
What is the value of y?

a b c

▪ 36

$$\frac{52}{65} = \frac{4}{x} = \frac{y}{z}$$

**28.** Terry and Pete take part in a marathon walk. Pete completes the walk in **6/16 of a day** and Terry completes the walk in **42/4 hours**.
How many **hours faster** was Pete?
*Give the answer as a mixed number fraction.*

a b c

▪ 1 1/2

**29.** At bingo, Millie wins 1/3 of the total prize money, Nora wins 16/51 and Albert wins the rest. What fraction of the money did **Albert** win?
*Give your answer in the simplest form.*

a b c

▪ 6/17

**Level 4:** *cont.*

**30.** Billy is thinking of a fraction equivalent to 3/21.
**a** The numerator is an odd number **between 20 and**
**b** **30**.
**c** The denominator is **greater than 200**.
What is Billy's fraction?

▪ 29/203

# Convert between mixed numbers and improper fractions

**Competency:** Convert between proper and improper fractions and mixed numbers.

**Quick Search Ref:** 10017

Correct: Correct.     Wrong: Incorrect, try again.     Open: Thank you.

**Level 1:** Understanding - Terminology and the value of digits in mixed number and improper fractions.

**✿ Required:** 7/10          **✿ Student Navigation:** on          **✿ Randomised:** off

**1.** Which of the values is a **mixed number fraction**?

■ 4 3/4  ■ 5/6  ■ 45/30

1/3

**2.** Which of the values is an **improper fraction**?

■ 7/8  ■ 2 4/5  ■ 13/8

1/3

**3.** How many parts make a whole in 3 7/8?

■ 3  ■ 7  ■ 8

1/3

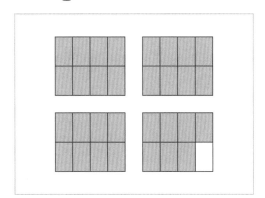

**4.** How many wholes are in 2 1/3?

■ 2  ■ 1  ■ 3

1/3

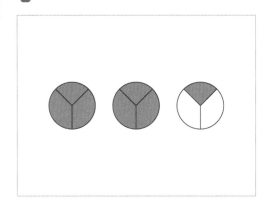

**5.** What is the denominator when 14/9 is written as a mixed number fraction?

■ 9

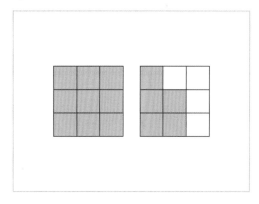

**6.** Give 19/5 as a mixed number fraction.

■ 3 4/5

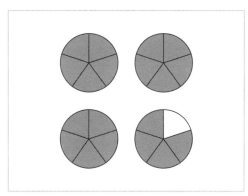

**7.** What is 4 3/7 as an improper fraction?

■ 31/7  ■ 28/7

1/2

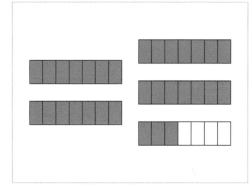

**Level 1:** *cont.*

**8.** What is 13/2 as a mixed number fraction?

a b c ▪ 6 1/2

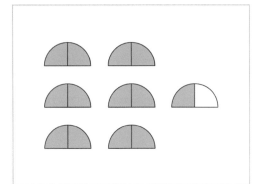

**9.** How many **parts** make a whole in 82/9?

a b c ▪ 9

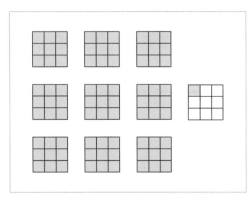

**10.** How many wholes are in 6 1/4?

a b c ▪ 6

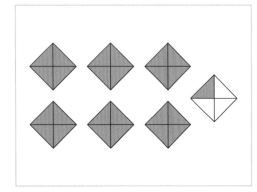

**Level 2:** Fluency - Recognising and converting between mixed number and improper fractions; simplest form.

❋ **Required:** 7/10   ❋ **Student Navigation:** on
❋ **Randomised:** off

**11.** What is the fraction represented by the shaded squares?
a b c *Give the answer as a mixed number fraction in its simplest form.*

▪ 3 5/8

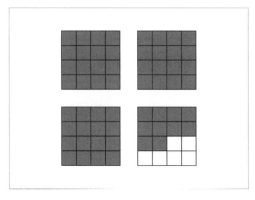

**12.** Convert 3 5/62 to an improper fraction.

a b c ▪ 191/62

**13.** Convert 204/22 to a mixed number fraction in its simplest form.

a b c ▪ 9 3/11

**14.** A school canteen serves pies that are cut into 34 equal pieces. One day, they sell 178 pieces of pie. What is this as a mixed number fraction in its simplest form?

a b c

▪ 5 4/17

**15.** Give 481/15 as a mixed number fraction in its simplest form.

a b c ▪ 32 1/15

**16.** After giving all of her students a pencil, Miss Prestage has 182 pencils left.
a b c If pencils come in packs of 124, how many packs does Miss Prestage have left?
*Give your answer as a mixed number fraction in its simplest form.*

▪ 1 29/62

**17.** Bottles of orange juice are delivered to a shop in packs of 45.
a b c If the shop sells 8 full packs and 18 bottles of orange juice, what is the amount as an improper fraction?

▪ 42/5

Level 2: *cont.*

**18.** Convert 5 34/50 to an improper fraction.

 ▪ 142/25 ▪ 284/50

**19.** What is 62/54 as a mixed number in its simplest form?

▪ 1 4/27

**20.** Give the fraction represented by the shaded parts as an improper fraction.

▪ 19/8

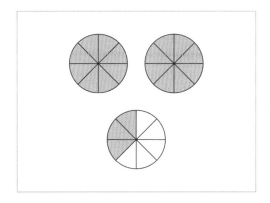

Level 3: Reasoning - Comparing mixed number and improper fractions.

✱ **Required:** 5/5  ✱ **Student Navigation:** on
✱ **Randomised:** off

**21.** Which symbol makes the following statement true?
4/5 + 13/5 ___ 4/13 + 5/13

1/3  ▪ < ▪ = ▪ >

**22.** What is the missing numerator in the equation?

▪ 18

$$\frac{8}{11} + \frac{?}{11} = 2\frac{4}{11}$$

**23.** Arrange the fractions in ascending order (smallest first).

▪ 73/6 ▪ 12 1/4 ▪ 12 1/3 ▪ 12 1/2 ▪ 64/5

**24.** What fraction is missing from the following equation?
8 - ___ = 48/9
*Give your answer as a mixed number fraction in its simplest form.*

▪ 2 2/3

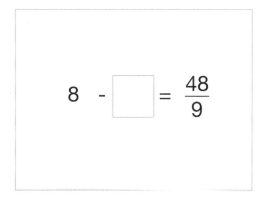

**25.** Which number is greater 31 2/3 or 199/7? Explain your answer.

Level 4: Problem Solving - Converting between and adding mixed number and improper fractions in context.

✱ **Required:** 5/5  ✱ **Student Navigation:** on
✱ **Randomised:** off

**26.** Convert the mixed number fractions to improper fractions and the improper fractions to mixed number fractions. Look up the answers in their simplest form in the table.
What **6-letter word** does the code spell?
9 2/5
7/3
2 3/10
3/2
7/3
5 9/13

▪ vertex ▪ vertex.

| A | $3\frac{2}{5}$ | H | $\frac{7}{4}$ | O | $2\frac{1}{4}$ | U | $\frac{23}{5}$ |
|---|---|---|---|---|---|---|---|
| B | $2\frac{4}{5}$ | I | $5\frac{1}{2}$ | P | $\frac{10}{3}$ | V | $\frac{47}{5}$ |
| C | $\frac{19}{12}$ | J | $4\frac{2}{3}$ | Q | $\frac{14}{11}$ | W | $\frac{13}{4}$ |
| D | $\frac{15}{7}$ | K | $\frac{16}{7}$ | R | $\frac{23}{10}$ | X | $\frac{74}{13}$ |
| E | $2\frac{1}{3}$ | L | $1\frac{7}{8}$ | S | $\frac{22}{10}$ | Y | $\frac{14}{9}$ |
| F | $\frac{18}{7}$ | M | $2\frac{7}{8}$ | T | $1\frac{1}{2}$ | Z | $\frac{11}{8}$ |
| G | $\frac{13}{5}$ | N | $\frac{17}{6}$ | | | | |

**Level 4:** *cont.*

**27.** The fractions are arranged in descending order (largest first). What is the missing denominator?

1
2
3

▪ 15

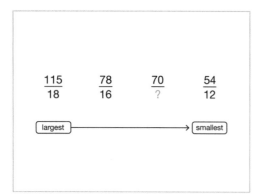

**28.** Brooke makes the fraction closest to 4 using **two** of her number cards.

a
b
c

What is her fraction written as a **mixed number fraction**?

▪ 4 1/5

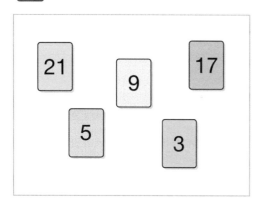

**29.** What is the product of *x*, *y* and *z*?

a
b
c

▪ 84

$$\frac{25}{x} = y\frac{z}{7}$$

**30.** Maisey is 12 7/8 years old.

a
b
c

Her Aunt Kelly is 84/5 years older than Maisey.
Maisey's Grandad is twice Aunt Kelly's age.
How old is Maisey's Grandad?
*Give your answer as a mixed number fraction in its simplest form.*

▪ 59 7/20

# Mathematics Y7

## Decimals and Percentages

### Equivalence

# Convert between fractions and decimals

**Competency:**    Convert between fractions and decimals.

**Quick Search Ref:**    10120

Correct: Correct.    Wrong: Incorrect. Try again.    Open: Thank you.

**Level 1:**    Understanding - Value of digits in fractions and decimals to calculate basic conversions.

✹ Required: 10/10        ✹ Student Navigation: on        ✹ Randomised: off

**1.** Select the **3** terms that can represent **parts of a whole.**

3/5  ■ multiple  ■ percentage  ■ square number  ■ fraction
    ■ decimal

**2.** How do you convert a fraction to a decimal?

1/3  ■ Divide the numerator by the denominator.
    ■ Divide the denominator by the numerator.
    ■ Multiply the numerator by the denominator.

**3.** When converting 7/8 to a decimal, what calculation do you use?

1/3  ■ 7 ÷ 8  ■ 8 ÷ 7  ■ 7 × 8

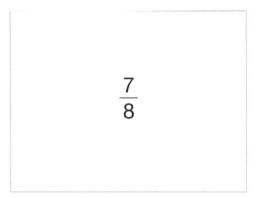

$$\frac{7}{8}$$

**4.** When converting from a decimal to a fraction, what determines the denominator?

1/4  ■ The last digit.
    ■ The column that represents the last digit.
    ■ The first digit.
    ■ The column that represents the first digit.

**5.** What is 0.17 as a fraction?

■ 17/100

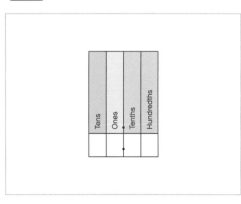

**6.** What is 3/5 as a decimal?

■ 0.6

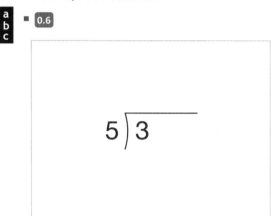

$5\overline{)3}$

**7.** Express 81/100 as a decimal.

■ 0.81

**8.** What is 0.183 as a fraction?

■ 183/1,000  ■ 183/1000

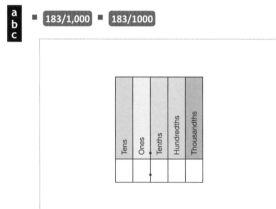

**9.** What is 1/5 as a decimal?

■ 0.2

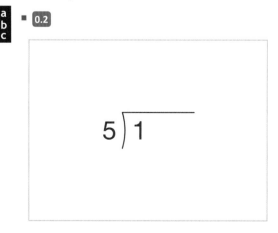

$5\overline{)1}$

**Level 1:** *cont.*

**10.** Express 1/8 as a decimal.

 ▪ 0.125

---

**Level 2:** Fluency - Recognising and converting between fractions and decimals (including mixed number fractions).

✹ **Required:** 7/10     ✹ **Student Navigation:** on
✹ **Randomised:** off

**11.** When converting a **mixed number fraction** to a decimal:

1/2

▪ convert the fraction part to a decimal and add the decimal to the whole number.

▪ convert to an improper fraction then divide the numerator by the denominator.

$$3 \frac{2}{10}$$

**12.** Express 4 3/10 as a decimal.

 ▪ 4.3

**13.** Calculate 3/8 as a decimal.

abc ▪ 0.375

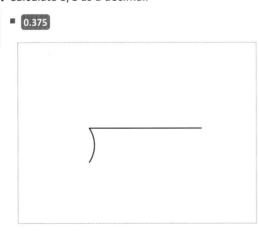

**14.** Calculate 0.84 as a fraction.
*Give your answer in its simplest form.*

abc ▪ 21/25

**15.** Richard cooks 5/16 kg of rice. What is this as a decimal?
*Include the units kg (kilograms) in your answer.*

abc ▪ 0.3125 kg  ▪ 0.3125 kilograms

**16.** Carlos, Eve and Asya have a packet of raisins each.
Carlos eats 0.85 of his raisins.
Eve eats 41/50 of her raisins.
Asya eats 4/5 of her raisins.
1/3  Who has eaten the most raisins?

▪ Carlos  ▪ Eve  ▪ Asya

**17.** What is the missing numerator?

abc ▪ 9

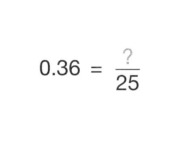

$$0.36 = \frac{?}{25}$$

**18.** What is 1 2/5 as a decimal?

abc ▪ 1.4

**19.** Calculate 3/20 as a decimal.

abc ▪ 0.15

**20.** What is 0.234 as a fraction?
*Give your answer in its simplest form.*

abc ▪ 117/500

**Level 3:** Reasoning - Comparing, ordering and reasoning with fractions and decimals.

✿ **Required:** 5/5    ✿ **Student Navigation:** on
✿ **Randomised:** off

**21.** Use the table to help you calculate 4/9 as a
**a** **decimal** to **3 decimal places** (3 d.p.).
**b**
**c** ▪ [0.444]

| $\frac{1}{9}$ | $0.\dot{1}$ |
|---|---|
| $\frac{2}{9}$ | $0.\dot{2}$ |
| $\frac{3}{9}$ | $0.\dot{3}$ |

**22.** Lily converts 7/20 to a decimal by dividing 7 by 20.
**a** Hugo says he can calculate this without dividing. Is
**b** Hugo correct? Explain your answer.
**c**

**23.** Which symbol makes the following statement
☐ true?
☒ 0.627 ___ 5/8
☐
1/3  ▪ <  ▪ =  ▪ [>]

**24.** Arrange the numbers in ascending order (smallest
↑ first).
↓
▪ [0.001]  ▪ [8/100]  ▪ [0.095]  ▪ [15/15]  ▪ [1.42]

**25.** Select the fraction that has a value closest to 0.25.
☐ ▪ 4/9  ▪ 1/5  ▪ 2/7  ▪ [4/15]
☒
☐
1/4

**Level 4:** Problem Solving - Converting between fractions (including improper and mixed number fractions) and decimals.

✿ **Required:** 5/5    ✿ **Student Navigation:** on
✿ **Randomised:** off

**26.** Two friends have **18 1/8** litres of orange juice to
**1** sell at the summer fair. At the end of the day they
**2** have **3.46** litres left. How many litres did they sell?
**3** *Give your answer as a decimal.*

▪ [14.665]

**27.** The values are shown in descending order (largest
**1** first). What is the missing numerator?
**2**
**3** ▪ [1]

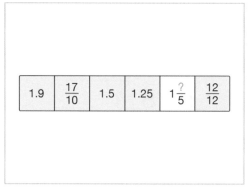

**28.** Can you crack the code? Convert the decimals to
**a** fractions and the fractions to decimals. Look up
**b** the values in the table and convert them to letters.
**c** 8 4/8 =
9 1/5 =
3/12 =
3.04 =
9 20/100 =
2 3/30 =
*All answers are in the simplest form.*

▪ [reflex.]  ▪ [reflex]

| A | 1.06 | H | 2.5 | O | $4\frac{1}{5}$ | U | $4\frac{3}{4}$ |
|---|---|---|---|---|---|---|---|
| B | $3\frac{1}{4}$ | I | 0.34 | P | 21.2 | V | 6.2 |
| C | 1.6 | J | $\frac{6}{2}$ | Q | 10.06 | W | 9.002 |
| D | 0.342 | K | 0.16 | R | 8.5 | X | 2.1 |
| E | 9.2 | L | $3\frac{1}{25}$ | S | 0.0342 | Y | 3.42 |
| F | 0.25 | M | $9\frac{2}{5}$ | T | $\frac{16}{17}$ | Z | $1\frac{8}{10}$ |
| G | 2.9 | N | 7.0 | | | | |

**29.** Add two side-by-side fractions to find the answer
**a** to the box above. What numerator is missing from
**b** the highlighted box?
**c**

▪ [5]

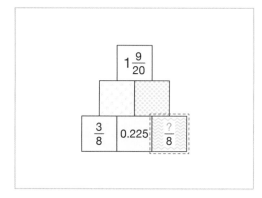

**30.** What denominator is missing from the calculation?

▪ 24

$$\frac{3}{8} + \frac{12}{?} = 0.875$$

# Mathematics

**Y7**

## Properties of Number

Multiples and Factors

Prime Numbers

Powers and Roots

Squares and Cubes

# Find the highest common factor using prime factor decomposition

**Competency:** Use the concepts and vocabulary of prime numbers, factors (or divisors), common factors, common multiples, highest common factor, lowest common multiple, prime factorisation, including using product notation and the unique factorisation property.

**Quick Search Ref:** 10197

Correct: Correct.   Wrong: Incorrect, try again.   Open: Thank you.

**Level 1:** Understanding - Review factors, HCF and prime factor decomposition.

✹ **Required:** 7/10   ✹ **Student Navigation:** on   ✹ **Randomised:** off

**1.** What is a factor?

1/4
- ■ A whole number which divides exactly into another whole number.
- ■ A number that can only be divided by itself and 1.
- ■ A number that can be divided exactly by whole numbers other than itself and 1.
- ■ A number which is multiplied by itself.

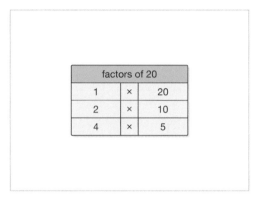

| factors of 20 | | |
|---|---|---|
| 1 | × | 20 |
| 2 | × | 10 |
| 4 | × | 5 |

**2.** The table shows the factors of 48 and 60. What is the highest common factor of 48 and 60?

  ■ 12

| factors of 48 | | | | factors of 60 | | |
|---|---|---|---|---|---|---|
| 1 | × | 48 | | 1 | × | 60 |
| 2 | × | 24 | | 2 | × | 30 |
| 3 | × | 16 | | 3 | × | 20 |
| 4 | × | 12 | | 4 | × | 15 |
| 6 | × | 8 | | 5 | × | 12 |
| | | | | 6 | × | 10 |

**3.** Fill in the missing factors of 56 and of 84.
What is the highest common factor of 56 and 84?

  ■ 28

| factors of 56 | | | | factors of 84 | | |
|---|---|---|---|---|---|---|
| 1 | × | 56 | | 1 | × | 84 |
| 2 | × | | | 2 | × | |
| 4 | × | 14 | | 3 | × | 28 |
| | × | | | | × | 21 |
| | | | | | × | |
| | | | | 7 | × | 12 |

**4.** What is a prime factor?

1/4
- ■ A whole number which divides exactly into another number.
- ■ A number that can only be divided by itself and 1.
- ■ A number that can be divided exactly by whole numbers other than itself and 1.
- ■ A factor that is also a prime number.

**5.** What are the two missing prime factors of 210?

2/6
■ 1 ■ 2 ■ 3 ■ 5 ■ 7 ■ 11

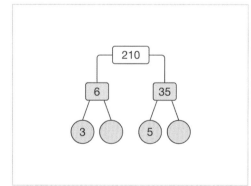

**Level 1:** *cont.*

**6.** Which expression shows 60 written as a product of its prime factors?

■ 2 × 3 × 5  ■ 2, 2, 3, 5  ■ 6 × 10  ■ 2 × 2 × 3 × 5

1/4

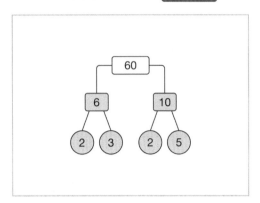

**7.** Look at the Venn diagram and select which numbers are the prime factors of both 350 and 60.

■ 3 and 7  ■ 2 and 3  ■ 2 and 5  ■ 5 and 7

1/4

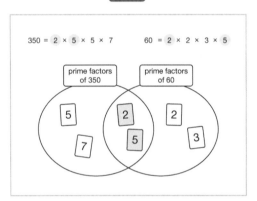

**8.** Fill in the missing factors of 48 and of 72. Use the lists to find the highest common factor of 48 and 72.

■ 24

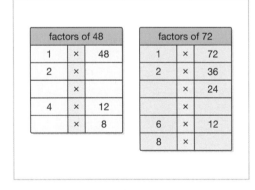

**9.** What is 108 as a product of its prime factors?
*Give your prime factors in ascending order.*

a
b
c

■ 2 * 2 * 3 * 3 * 3  ■ 2 × 2 × 3 × 3 × 3  ■ 2 x 2 x 3 x 3 x 3

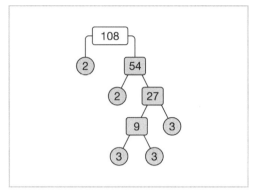

**10.** Look at the Venn diagram and select the prime factors of both 126 and 735.

■ 2 and 5  ■ 5 and 7  ■ 2 and 3  ■ 3 and 7

1/4

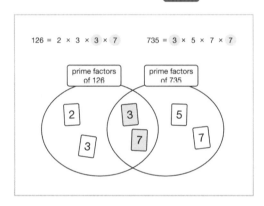

**Level 2:** Fluency - Find the HCF using prime factor decomposition.

❋ **Required:** 7/10   ❋ **Student Navigation:** on
❋ **Randomised:** off

**11.** The Venn diagram shows the prime factors of 120 and 144.
The highest common factor is the product of all the prime factors in the intersection.
What is the highest common factor of 120 and 144?

■ 24

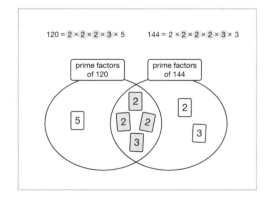

**Level 2:** *cont.*

**12.** Use the Venn diagram to find the highest common factor of 112 and 280.

1
2
3   ▪ 56

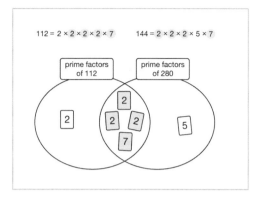

**13.** Find the highest common factor of 126 and 210.

1
2
3   ▪ 42

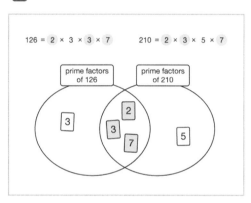

**14.** Find the highest common factor of 156 and 390.

1
2
3   ▪ 78

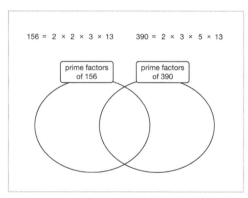

**15.** Find the highest common factor of 84 and 196.

1
2
3   ▪ 28

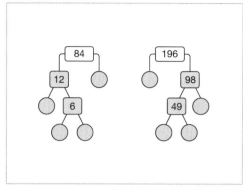

**16.** Find the highest common factor of 90 and 126.

1
2
3   ▪ 18

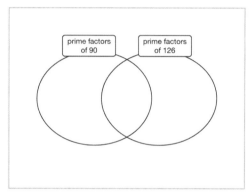

**17.** Find the highest common factor of 78 and 130.

a
b
c   ▪ 26

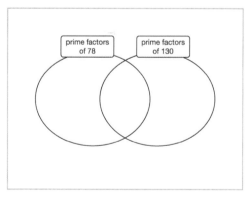

**18.** Complete the factor trees and find the highest common factor of 171 and 228.

1
2
3   ▪ 57

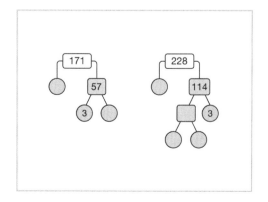

**Level 2:** *cont.*

**19.** Find the highest common factor of 108 and 180

  ▪ 36

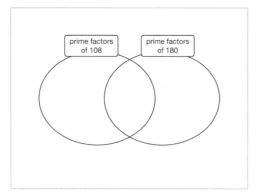

**20.** Find the highest common factor of 175 and 245.

  ▪ 35

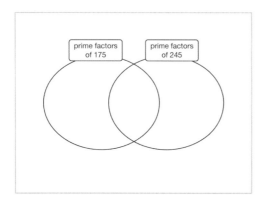

**Level 3:** Reasoning - Misconceptions with HCF and applying HCF.

✸ **Required:** 5/5   ✸ **Student Navigation:** on
✸ **Randomised:** off

**21.** Petra has calculated the highest common factor of 390 and 462 as 36.
Is Petra correct? Explain your answer.

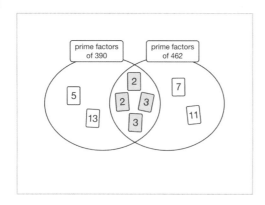

**22.** Mrs. Carter has 140 crayons and 196 colouring pencils. What is the most students Mrs. Carter can divide her equipment between so that every student receives the same amount of crayons and pencils?

▪ 28

**23.** Tayyab has calculated the highest common factor of 126 and 315 as 21.
Is Tayyab correct? Explain your answer.

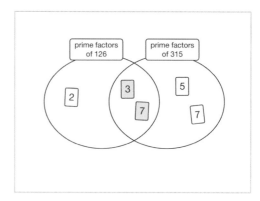

**24.** Simplify the following fraction:
154/385

▪ 2/5

**25.** The highest common factor of 105 and 394 is 21.
What is the highest common factor of 210 and 394?

▪ 42

**Level 4:** Problem Solving - Applying HCF to multi-step problems.

✸ **Required:** 5/5   ✸ **Student Navigation:** on
✸ **Randomised:** off

**26.** A rugby match has an attendance of 1,540 including 924 adults.
What fraction of the crowd are children? Give your answer in its simplest form.

▪ 2/5

**27.** Find the highest common factor of 420, 462 and 546.

▪ 42

**28.** Two numbers have a product of 1,350 and a highest common factor of 15.
90 is not one of the numbers. What is the difference between the two numbers?

▪ 15

**29.** Find the highest common factor of each pair of numbers and arrange them in ascending order.

▪ 266 and 456   ▪ 336 and 378   ▪ 336 and 432

**30.** Amir is tiling a wall that is 2.94 metres high and 4.2 metres wide.
Amir uses the largest square tile possible without cutting any tiles.
How many tiles does Amir use?

▪ 70

# Find the lowest common multiple using prime factor decomposition

**Competency:** Use the concepts and vocabulary of prime numbers, factors (or divisors), common factors, common multiples, highest common factor, lowest common multiple, prime factorisation, including using product notation and the unique factorisation property.

**Quick Search Ref:** 10287

Correct: Correct.    Wrong: Incorrect, try again.    Open: Thank you.

**Level 1:** Understanding - Review factors, LCM and prime factor decomposition.

✸ **Required:** 7/10    ✸ **Student Navigation:** on    ✸ **Randomised:** off

**1.** What is a multiple?

1/4
- ■ A whole number which divides exactly into another whole number.
- ■ A number that can only be divided by itself and 1.
- ■ The result of multiplying a number by an integer.
- ■ A number which is multiplied by itself.

**2.** The image shows multiples of 12 and multiples of 18 with the common multiples circled.
What is the lowest common multiple of 12 and 18?

- ■ 36

| multiples of | |
|---|---|
| 12 | 18 |
| 12 | 18 |
| 24 | (36) |
| (36) | 54 |
| 48 | (72) |
| 60 | 90 |
| (72) | (108) |
| 84 | 126 |
| 96 | |
| (108) | |
| 120 | |

**3.** Continue the lists of multiples of 24 and 42.
Use your lists to find the lowest common multiple of 24 and 42.

- ■ 168

| multiples of | |
|---|---|
| 24 | 42 |
| 24 | 42 |
| 48 | 84 |
| 72 | 126 |

**4.** What is a prime factor?

1/4
- ■ A whole number which divides exactly into another number.
- ■ A number that can only be divided by itself and 1.
- ■ A number that can be divided exactly by whole numbers other than itself and 1.
- ■ A factor that is a prime number.

**5.** What are the two missing prime factors of 330?

2/6
- ■ 1 ■ 2 ■ 3 ■ 5 ■ 7 ■ 11

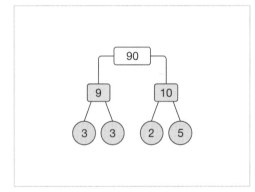

**6.** Which expression shows 90 written as a product of its prime factors?

1/4
- ■ 2 × 3 × 5 ■ 2, 3, 3, 5 ■ 9 × 10 ■ 2 × 3 × 3 × 5

**7.** Look at the Venn diagram and select which are the prime factors of both 294 and 315.

1/4
- ■ 2 and 5 ■ 2 and 7 ■ 3 and 7 ■ 3 and 5

294 = 2 × 3 × 7 × 7        315 = 3 × 2 × 5 × 7

prime factors of 294    prime factors of 315
2    3    3
7    7    5

**Level 1:** *cont.*

**8.** Complete the lists of multiples of 21 and 35.
Use your lists to find the lowest common multiple of 21 and 35.

■ 105

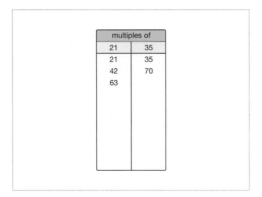

**9.** What is 132 as a product of its prime factors?
*Give your prime factors in ascending order.*

■ 2 x 2 x 3 x 11  ■ 2 × 2 × 3 × 11  ■ 2 * 2 * 3 * 11

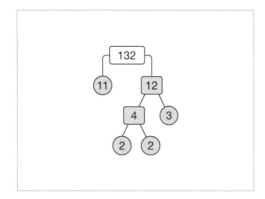

**10.** Look at the Venn diagram and select which are the prime factors of both 126 and 735.

■ 2 and 5  ■ 5 and 7  ■ 2 and 3  ■ 3 and 7

1/4

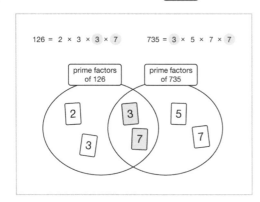

**Level 2:** Fluency - Find the LCM using prime factor decomposition.

✸ **Required: 7/10**  ✸ **Student Navigation: on**
✸ **Randomised: off**

**11.** The Venn diagram shows the prime factors of 24 and 30.
The lowest common multiple is the product of all prime factors in the Venn diagram.
What is the lowest common multiple of 24 and 30?

■ 120

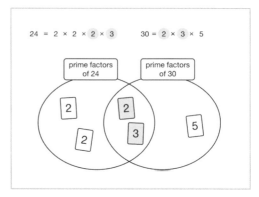

**12.** Use the Venn diagram to find the lowest common multiple of 45 and 70.

■ 630

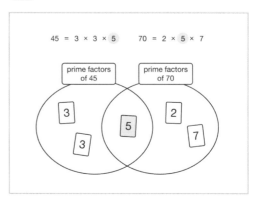

**13.** Find the lowest common multiple of 60 and 84.

■ 420

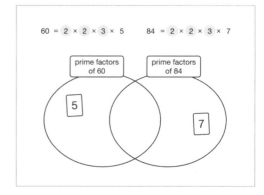

**14.** Find the lowest common multiple of 54 and 90.

 ▪ 270

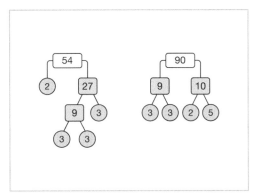

**15.** Find the lowest common multiple of 28 and 84.

 ▪ 84

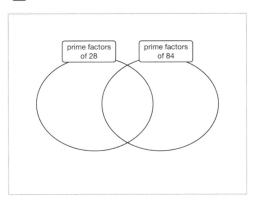

**16.** Find the lowest common multiple of 70 and 105.

 ▪ 210

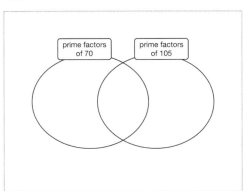

**17.** Find the lowest common multiple of 147 and 245.

 ▪ 735

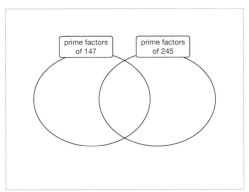

**18.** Find the lowest common multiple of 42 and 210.

 ▪ 210

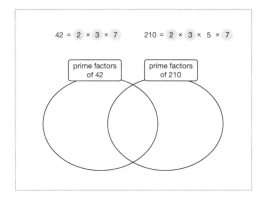

**19.** Find the lowest common multiple of 112 and 140.

 ▪ 560

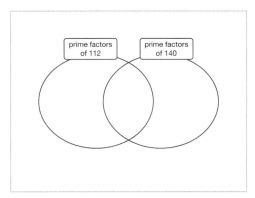

**20.** Find the lowest common multiple of 75 and 105.

▪ 525

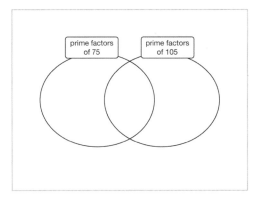

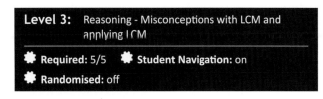

**Level 3:** Reasoning - Misconceptions with LCM and applying LCM

�')  **Required:** 5/5    🌺 **Student Navigation:** on
🌺 **Randomised:** off

**21.** Shola has calculated the lowest common multiple
    a    of 30 and 42 as 1,260.
    b
    c    Is Shola correct? Explain your answer.

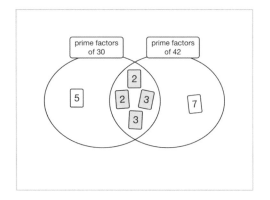

**22.** Mercury takes approximately 3 months to orbit
    1    the Sun and Venus takes approximately 8 months.
    2    If Mercury, Venus and Earth are all in a straight line
    3    with the Sun, how many years will it take the three
         planets to be in the same positions again?

    ▪ 2

**23.** The lowest common multiple of 51 and 85 is 255.
    1    What is the lowest common multiple of 102 and
    2    85?
    3

    ▪ 510

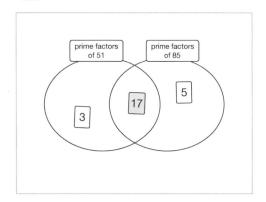

**24.** What is the lowest common multiple of 15, 20 and
    1    25?
    2
    3    ▪ 300

**25.** Skye says, "The product of two numbers is equal
    a    to the product of their lowest common multiple
    b    and their highest common factor".
    c    Is Skye correct? Explain your answer.

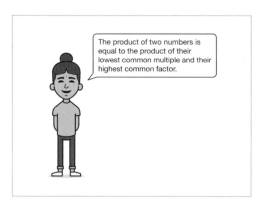

**Level 4:** Problem Solving - Multi-step problems
    applying LCM.

🌺 **Required:** 5/5    🌺 **Student Navigation:** on
🌺 **Randomised:** off

**26.** A bus leaves Sumtown station every 18 minutes
    1    and a train leaves every 14 minutes.
    2    A bus and train both leave at 9:00.
    3    How many more times before 17:00 will a bus and
         a train leave at the same time?

    ▪ 3

**27.** Calculate 13/84 - 7/48.
    a    Give your answer as a fraction in its simplest form.
    b
    c    ▪ 1/112

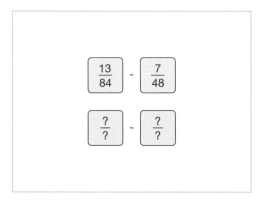

**28.** What is the lowest common multiple of 28, 30 and
    1    42?
    2
    3    ▪ 420

**29.** Find the lowest common multiple of each pair of
    ↑    numbers and arrange them in ascending order.
    ↓
    ▪ 28 and 77.  ▪ 45 and 105.  ▪ 81 and 108.

**30.** The highest common factor of two numbers is 14.
    1    The lowest common multiple of the same two
    2    numbers is 210.
    3    Both numbers are less than 100.
         What is the difference between the two numbers?

    ▪ 28

# Find the prime factor decomposition of a number

**Competency:** Use the concepts and vocabulary of prime numbers, factors (or divisors), common factors and highest common factor (HCF).

**Quick Search Ref:** 10123

Correct: Correct.   Wrong: Incorrect, try again.   Open: Thank you.

**Level 1:** Understanding - Definitions, testing divisibility, missing numbers on a factor tree.

✷ **Required:** 7/10    ✷ **Student Navigation:** on    ✷ **Randomised:** off

**1.** What is the product of 5 and 2?

 ■ 7 ■ 52 ■ 3 ■ **10**

1/4

**2.** Does 7,628 divide by 2 without leaving a remainder?

 ■ **Yes** ■ No

1/2

**3.** What number is missing from the factor tree?

 ■ **2**

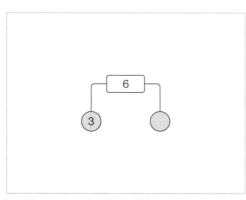

**4.** 92,185 is a multiple of 5. True or false?

 ■ **True** ■ False

1/2

**5.** Is 3 a factor of 7,812,791?

 ■ Yes ■ **No**

1/2

**6.** What number goes at the top of the factor tree?

 ■ **26**

**7.** What is a prime factor?

■ A whole number which exactly divides into another whole number.
■ A number that can only be divided by itself and 1.
■ A number that can be divided evenly by numbers other than itelf and 1.
■ **A factor that is a prime number.**

1/4

**8.** What is the missing **prime factor**?

 ■ **13**

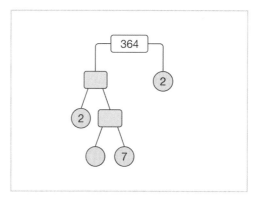

**9.** What is the missing **prime factor**?

 ■ **3**

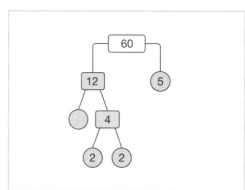

**Level 1: *cont.***

**10.** What number is missing from the factor tree?

$\frac{1}{2}$ = **7**

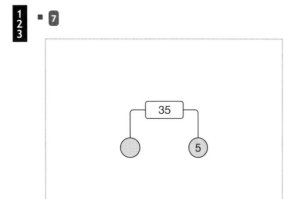

**Level 2:** Fluency - Writing numbers as a product of prime factors.

✿ **Required:** 7/10   ✿ **Student Navigation:** on
✿ **Randomised:** off

**11.** Sean is asked to write 30 as a product of its prime factors. What is the correct answer?

☐☒☐   ■ **2, 3, 5**  ■ **2 + 3 + 5**  ■ **2 × 3 × 5 × 6**  ■ **2 × 3 × 5**

1/4

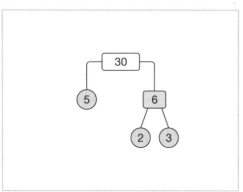

**12.** Write 26 as a product of its prime factors.

a b c   ■ **2 × 13**  ■ **13 x 2**  ■ **13 × 2**  ■ **2*13**  ■ **13*2**
■ **2 x 13**

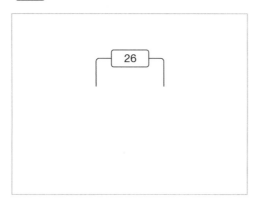

**13.** Complete the factor tree and write 36 as a product
a of its prime factors.
b Write the factors in **ascending** order.
c

■ **2 × 2 × 3 × 3**  ■ **2 x 2 x 3 x 3**  ■ **2*2*3*3**

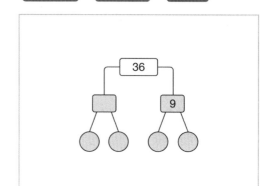

**14.** Write 42 as a product of its prime factors.
a Write the factors in **ascending** order.
b
c   ■ **2 x 3 x 7**  ■ **2*3*7**  ■ **2 × 3 × 7**

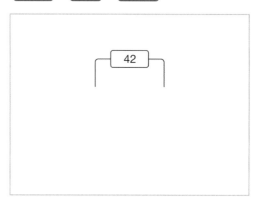

**15.** Write 128 as a product of its prime factors.

a   ■ **2 x 2 x 2 x 2 x 2 x 2 x 2**  ■ **2*2*2*2*2*2*2**
b
c   ■ **2 × 2 × 2 × 2 × 2 × 2 × 2**

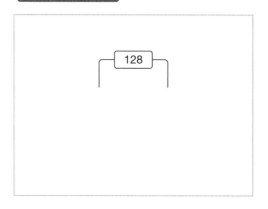

**Level 2: cont.**

**16.** If you used the cards shown to write 132 as a product of its prime factors, which three cards will you have **left over**?

3/5  ■ 2  ■ **3**  ■ **5**  ■ **7**  ■ 11

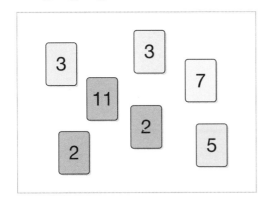

**17.** Write 1,092 as a product of its prime factors? Write the factors in **ascending** order.
a b c

■ **2 x 2 x 3 x 7 x 13**  ■ **2*2*3*7*13**  ■ **2 × 2 × 3 × 7 × 13**

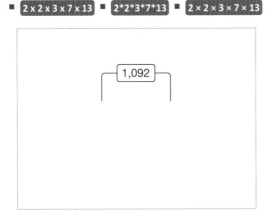

**18.** Write 28 as a product of its prime factors. Write the factors in **ascending** order.
a b c

■ **2 × 2 × 7**  ■ **2*2*7**  ■ **2 x 2 x 7**

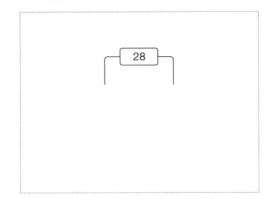

**19.** Write 660 as a product of prime factors. Write the factors in **ascending** order.
a b c

■ **2*2*3*5*11**  ■ **2 × 2 × 3 × 5 × 11**  ■ **2 x 2 x 3 x 5 x 11**

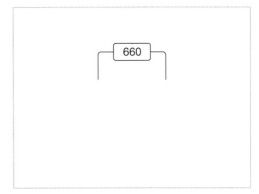

**20.** If you use the digit-cards to write 252 as a product of its prime factors, which three cards will you have **left over**?

3/5  ■ **2**  ■ 3  ■ **5**  ■ 7  ■ **11**

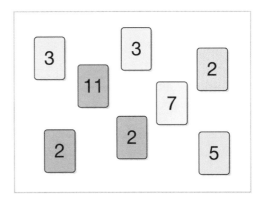

**Level 3:**  Reasoning - Patterns with products of prime factors.

✱ **Required:** 5/7   ✱ **Student Navigation:** on
✱ **Randomised:** off

**21.** Given that the prime factor decomposition of 72 is **2 × 2 × 2 × 3 × 3**. What is 720 written as a product of its prime factors?
a b c
Write the factors in ascending order.

■ **2 × 2 × 2 × 2 × 3 × 3 × 5**  ■ **2*2*2*2*3*3*5**
■ **2 x 2 x 2 x 2 x 3 x 3 x 5**

**Level 3:** *cont.*

**22.** Alicia, Brad and Chantelle each complete a prime factor decomposition for the same number. Who completed it **incorrectly**?

1/3   ■ **Alicia**   ■ **Brad**   ■ **Chantelle**

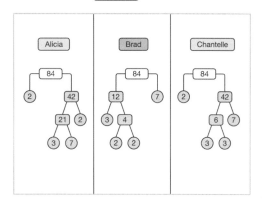

**23.** Malik says, "The larger a number is, the more prime factors it will have".
Is Malik correct? Explain your answer.

**24.** How do you know that 6 is not a factor of 572,126? Explain your reasoning.

**25.** Jamal writes 462 and 2,805 as products of their prime factors.
Which **two numbers** are in both of the products?

2/6   ■ 2   ■ **3**   ■ 5   ■ 7   ■ **11**   ■ 17

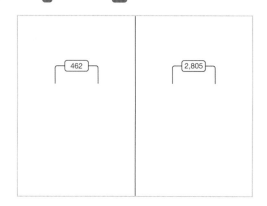

**26.** Shona says, "Any factor of a number is either a prime factor or a product of its prime factors".
Is Shona correct? Explain your answer.

**27.** Andrew has drawn a factor tree to help him work out 70 as a product of its prime factors. How many different ways could Andrew complete the factor tree without changing the branches in any way?

■ **6**

**Level 4:**   Problem Solving - Problems involving the product of prime factors.

✿ **Required:** 5/5    ✿ **Student Navigation:** on
✿ **Randomised:** off

**28.** To write 120 as a product of prime factors, Zain has started a tree diagram with the factors 10 and 12.
Which **two** statements are **true**?

2/5
- He should have started off with 2 and 60.
- **The next set of factor pairs could be 2, 5 and 3, 4.**
- The next set of factor pairs could be 5, 5 and 2, 6.
- The prime factors of 120 are 2, 3, 4 and 5.
- **The prime factors of 120 are 2, 3 and 5.**

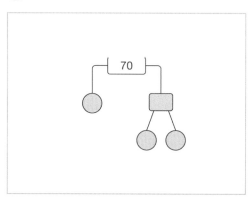

**29.**  Fazel arranges the digits 2, 4 and 6 to make a three-digit number and then writes the number as a product of its prime factors.
Which way of arranging the digits would give Fazel the largest number of prime factors in the product?

- 624

**30.** If you write the fraction shown as simply as possible, what will the numerator be?

- 2

$$\frac{1,092}{2,730}$$

**31.** Joel is thinking of a three-digit number which is the product of two prime numbers.
What is the smallest number Joel could be thinking of?

- 106

**32.** How many factors does 910 have?

 - 16

# Recognise and use powers up to the power of 5 and corresponding roots

**Competency:** Use integer powers and associated real roots (square, cube and higher), recognise powers of 2, 3, 4, 5 and distinguish between exact representations of roots and their decimal approximations.

**Quick Search Ref:** 10112

Correct: Correct.     Wrong: Incorrect, try again.     Open: Thank you.

**Level 1:** Understanding - Using correct terminology and symbolic representation. Calculating roots and powers up to 10⁵.

✿ **Required:** 7/10     ✿ **Student Navigation:** on     ✿ **Randomised:** off

---

1. Select the option that shows 3 to the power of 4.

   ▪ 3 x 4  ▪ 4³  ▪ $3^4$  ▪ ⁴√3  ▪ ³√4

   1/5

   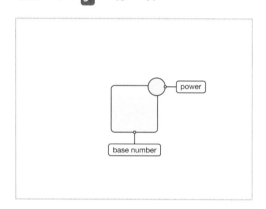

2. Which calculation would you use to work out 6⁴?

   ▪ 6 x 4  ▪ 6 + 6 + 6 + 6  ▪ 6 x 6 x 6 x 6  ▪ 4 x 4 x 4 x 4 x 4 x 4

   1/4

3. Work out 4 to the power of 5.

   ▪ 1,024  ▪ 1024

4. Calculate 10⁵.

   ▪ 100,000  ▪ 100000

5. The 5th root (⁵√) of __ = 2.

   ▪ 32

   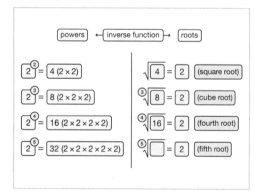

6. Select the calculations that would help you to find ⁵√1,024?

   ▪ 1,024 ÷ 5  ▪ 1,024 x 1024 x 1024 x 1024 x 1024
   ▪ 4 x 4 x 4 x 4 x 4  ▪ 1,024 x 5

   1/4

   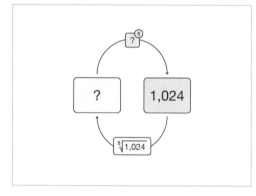

7. ⁴√81 = ?

   ▪ 3

   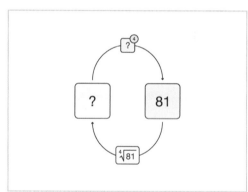

8. What is 8⁴?

   ▪ 4096  ▪ 4,096

---

**9.** Work out $^5\sqrt{3,125}$.

**abc** ▪ **5**

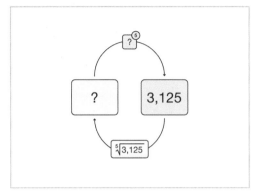

**10.** Calculate $^4\sqrt{16}$.

**abc** ▪ **2**

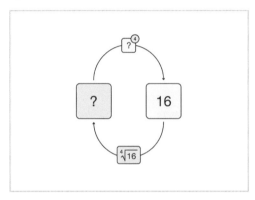

**Level 2:** Fluency - Calculating roots and powers up to 5, including negative integers.

✱ **Required:** 7/10    ✱ **Student Navigation:** on
✱ **Randomised:** off

**11.** What is the missing power?

**123** ▪ **5**

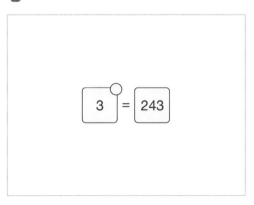

**12.** Work out $3^5 + 2^4$.

**123** ▪ **259**

**13.** Calculate $2^5 - {}^4\sqrt{625}$.

**123** ▪ **27**

**14.** What is the missing digit?

**abc** ▪ **5**

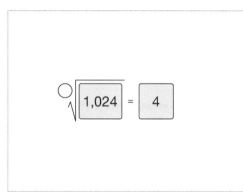

**15.** Calculate $(-10)^4$.

**abc** ▪ **10,000** ▪ **10000**

**16.** What is $^5\sqrt{243} \times 4^4$?

**abc** ▪ **768**

**17.** Arrange these expressions in descending order, (largest first).

↑↓ ▪ $(-2)^4 - (-2)^3$ ▪ $(-2)^4$ ▪ $(-2)^5 + (-2)^4$ ▪ $(-2)^5$

**18.** What is the missing power?

**123** ▪ **4**

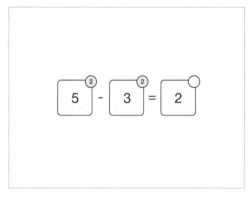

**19.** Find $5^4 + 4^5$.

**abc** ▪ **1649** ▪ **1,649**

**20.** What is $(-2)^5$?

**abc** ▪ **-32**

**Level 3:** Reasoning - Understanding the properties of powers and roots.

✹ **Required:** 5/5  ✹ **Student Navigation:** on
✹ **Randomised:** off

**Level 4:** Problem Solving - Using powers and roots to solve complex problems.

✹ **Required:** 5/5  ✹ **Student Navigation:** on
✹ **Randomised:** off

**21.** Prove that $4^3$ is the same as $2^6$.

**22.** Given that $6^6 = 46,656$, select the calculation that you would use to find $6^5$.

1/4  ■ 46,656 x 6 =   ■ 46,656 + 6 =   ■ 46,656 ÷ 6 =

■ 46,656 - 6 =

**23.** Harry notices the pattern shown when working out powers. Using a similar pattern, calculate $6^0$.

 ▪ **1**

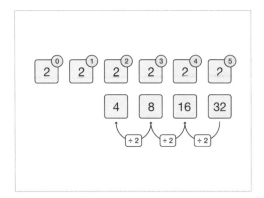

**24.** Use partitioning to calculate $\sqrt[5]{3,200,000}$.

▪ **20**

**25.** Explain why a power of 2 can never be a multiple of 10.

**26.** Find the value of $a$ when $a > 2$.

▪

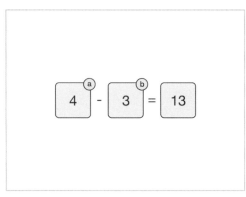

**27.** A super bacterium splits into four separate bacteria every 20 minutes. At 15:00 there was 1 bacterium. How many bacteria will there be at 16:40?

■ **1,024**   ■ **1024**

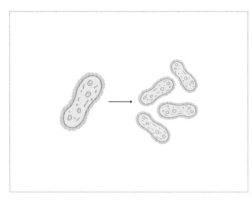

**28.** Using the digits 1 to 6 make this calculation true (you must use each digit only once).
What is the answer to the addition?

▪ **65**

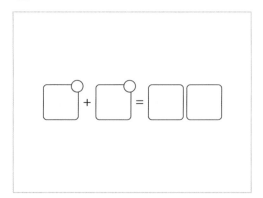

**29.** A piece of paper is 0.05 millimetres thick. Matias
folds the paper in half 5 times. How thick will
the folded piece of paper be?
*Include the units mm (millimetres) in your answer.*

a
b
c

- 1.6 millimetres   - 1.6 mm

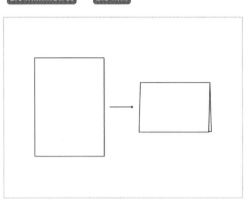

**30.** A crystal grows four times longer every year. If it is
0.1 mm long in 2017, in what year will its length
be 10.24 cm?

1
2
3

- 2022

# Understand cube numbers and cube roots

**Competency:** Use integer powers and associated roots (square, cube and higher).

**Quick Search Ref:** 10218

**Correct:** Correct.    **Wrong:** Incorrect, try again.    **Open:** Thank you.

**Level 1:** Understanding - Cube numbers from $1^3$ to $5^3$ and their corresponding roots.

✿ **Required:** 7/10    ✿ **Student Navigation:** on    ✿ **Randomised:** off

**1.** A cube number is:

1/3
- The result of multiplying a number by 3.
- The result of adding a number to itself twice.
- The result of multiplying a whole number by itself then by itself again.

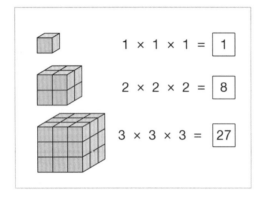

$1 \times 1 \times 1 = \boxed{1}$

$2 \times 2 \times 2 = \boxed{8}$

$3 \times 3 \times 3 = \boxed{27}$

**2.** $5^3$ means:

1/4
- $5 \times 3$  ■ $5+5+5$  ■ $5 \times 5 \times 5$  ■ $5 \times 5 \times 5 \times 5$

$$5^3$$

**3.** What is the value of $4^3$?

  ■ 64

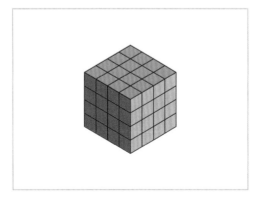

**4.** What is the 5 cubed?

 ■ 125

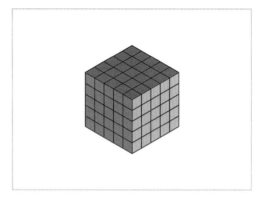

**5.** Which of the following is a cube number?

1/4
- ■ 1 ■ 65 ■ 9 ■ 30

**6.** What is the cube root of a number?

1/3
- The result of dividing the number by itself three times.
- A value which makes the number when multiplied by itself twice.
- The result of dividing the number by 3.

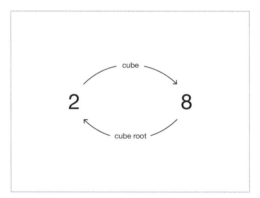

cube

2          8

cube root

**7.** How would you represent **the cube root of 27** in symbol form?

1/4
- ■ 3√27  ■ ³√27  ■ √27  ■ 27 ÷ 3

**Level 1:** *cont.*

**8.** What is the cube root of 27?

 ▪ **3**

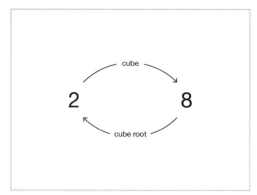

**9.** What is $\sqrt[3]{64}$?

 ▪ **4**

**10.** What is $1^3$?

 ▪ **1**

**Level 2:** Fluency - Fractions, cube numbers multiplied by powers of 10 and negative cube roots.

✿ **Required:** 7/10   ✿ **Student Navigation:** on
✿ **Randomised:** off

**11.** A cube is made up of 125 centimetre cubes. What is the side-length of the cube in centimetres?
*Include the units cm (centimetres) in your answer.*

▪ **5cm** ▪ **5centimeters** ▪ **5centimetres**

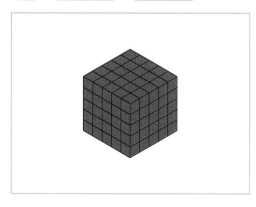

**12.** What fraction comes next in the following sequence?
1, 1/8, 1/27, ___

▪ **1/64** ▪ **1/4³**

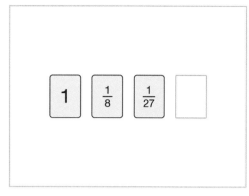

**13.** The cube root of 100 falls between which **two** integers?

▪ **4** ▪ **5** ▪ **33** ▪ **34**

2/4

**14.** What is $-2^3$?

▪ **-8**

**15.** What is the cube root of -27?

▪ **-3**

**16.** Calculate $0.5^3$.

▪ **0.125**

**17.** What is the cube root of 64/125?

▪ **4/5**

**18.** What is the cube root of 64,000?

▪ **40**

**19.** The cube root of which number is 0.3?

▪ **0.027**

**20.** If 27 × 64 = 1,728, what is the cube root of 1,728?

▪ **12**

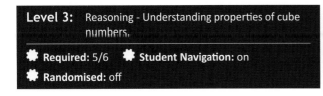

**Level 3:** Reasoning - Understanding properties of cube numbers.

✱ Required: 5/6 ✱ Student Navigation: on
✱ Randomised: off

**Level 4:** Problem solving - Application of cube numbers to solve problems.

✱ Required: 4/4 ✱ Student Navigation: on
✱ Randomised: off

**21.** Siobhan has a set of digit cards, which she is arranging to make a cube number. Siobhan thinks that it is impossible for the cube number to end in 1. Explain why she is wrong.

abc

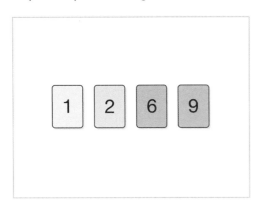

**22.** True or false? The difference between two consecutive cube numbers is **always** odd.

☐☒☐

■ Always true. ■ Sometimes true. ■ Never true.

1/3

**23.** Which digit represents the ones digit in the cube root of 148,877?

123

■ 3

**24.** How many four-digit cube numbers end in the digit 8?

123

■ 1

**25.** 3,375 is the product of two cube numbers. What is the cube root of 3,375?

123

■ 15

**26.** Ajay says that if you take a square number and multiply it by its square root, you will **always** end up with a cube number.
Prove that Ajay is correct.

abc

**27.** How many centimetre cubes are there in a metre cube?

abc

■ 1000000 ■ 1,000,000

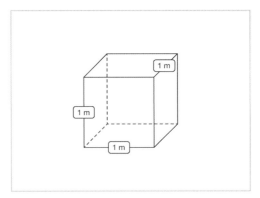

**28.** A cube is made up of 125 blue centimetre cubes. If the outside of the cube is painted red, how many of the centimetre cubes will be **all blue**?

123

■ 27

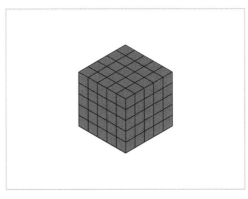

**29.** Ellie has a cube-shaped toy box with a volume of 126,000 cm³. Her building blocks each have a side length of 200 mm. How many blocks will fit inside the toy box?

123

■ 8

**30.**
a
b
c
A cube box with a side-length *s* is filled with 8 smaller cube boxes of sweets. What is the maximum **side-length** of each small box? Give your answer as an algebraic expression.

- s ÷ 2  - s/2

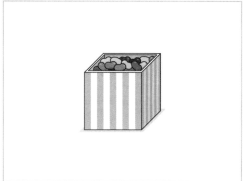

# Understand square numbers and their square roots

**Competency:** Use square numbers and their roots.

**Quick Search Ref:** 10281

**Correct:** Correct. **Wrong:** Incorrect, try again. **Open:** Thank you.

**Level 1:** Understanding - Square numbers from $1^2$ to $15^2$ and their corresponding square roots.

✹ **Required:** 7/10    ✹ **Student Navigation:** on    ✹ **Randomised:** off

1. A square number is:

   - The result of multiplying a number by 2.
   - The result of multiplying a whole number by itself.
   1/3   ■ The result of adding a number to itself.

2. $19^2$ means:

   - $19 \times 2$ ■ $19 + 19$ ■ $19 \times 19$ ■ $19 \times 19 \times 19$
   1/4

3. What is the value of $13^2$?

   - 169
   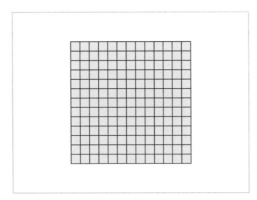

4. What is the square of 15?

   - 225
   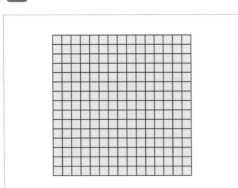

5. Which of the following is a square number?

   - 360 ■ 200 ■ 196 ■ 3.6
   1/4

6. What is the square root of a number?

   - The result of dividing the number by itself.
   - The result of dividing the number by 2.
   1/3 ■ The number that is multiplied by itself to make the square number.
   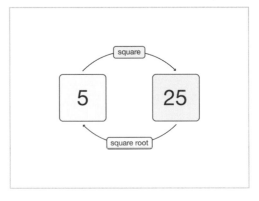

7. Which symbol represents the **square root of 9**?

   - $9 \div 2$ ■ $2\sqrt{9}$ ■ $\sqrt{9}$ ■ $^2\sqrt{9}$
   1/4

8. What is the positive square root of 144?

   - 12

9. Find the positive value of $\sqrt{121}$.

   - 11

10. What is the positive square root of 81?

    - 9

**Level 2:** Fluency - Extending to square larger numbers and decimal and use square roots.

✹ **Required:** 7/10   ✹ **Student Navigation:** on
✹ **Randomised:** off

11. Calculate $40^2$.
    - 1600 ■ 1,600

**12.** Calculate $30^2 - 9^2$

▪ 819

**13.** What is the answer to $\sqrt{36}$?
*Select 2 answers.*

▪ 6   ▪ 18   ▪ -18   ▪ -6

2/4

**14.** The square root of which number is 11?

▪ 121

**15.** Calculate $0.3^2$

▪ 0.09

**16.** What is the square of 0.5?

▪ 0.25

**17.** The square root of what number is 4?

▪ 16

**18.** Calculate $3^2 \times 4^2$.

▪ 144

**19.** What is $\sqrt{64} + \sqrt{36}$

▪ 14

**20.** A square has side length of 8 centimetres. Find the area of the square in $cm^2$.

▪ 64

**Level 3:** Fluency - Extending to look at fractions, known square numbers multiplied by powers of 10 and negative square roots.

❋ **Required:** 7/10   ❋ **Student Navigation:** on
❋ **Randomised:** off

**21.** Calculate $(-4)^2$.

▪ 16   ▪ +16

**22.** True or false? A number has only one square root.

▪ True   ▪ False

1/2

**23.** Find the positive square root of 90,000.

▪ 300

**24.** What is the square root of 1,600?

▪ 40

**25.** Calculate $0.07^2$.

▪ 0.0049

**26.** What is the square root of 1/4?

▪ 1/2   ▪ 1/8   ▪ 1/16

1/3

**27.** The square root of what number is 1.4?

▪ 1.96

**28.** If $49 \times 36 = 1,764$, what is the square root of 1,764?

▪ 42

**29.** A square has an area of 3,600 $cm^2$. Find the length of the side in centimetres.

▪ 60

**30.** Calculate $\sqrt{4} + \sqrt{16}$?

**Level 4:** Reasoning - using the properties of square numbers.

❋ **Required:** 6/8   ❋ **Student Navigation:** on
❋ **Randomised:** off

**31.** A, B and C are numbers.
A has 2 factors.
B has 3 factors.
C has 4 factors.

1/3

Which number is a square number?

▪ A   ▪ B   ▪ C

**Level 4: cont.**

**32.** George has a set of digit cards, which he is arranging to make a square number. Which digit **cannot** be placed in the ones column?

1/4    ▪ **3** ▪ 4 ▪ 5 ▪ 6

**33.** The difference between two consecutive square numbers will always be odd.

1/3    ▪ **Always true** ▪ Sometimes true ▪ Never true

**34.** Which **two digits** could represent the ones digit in the square root of 3,481?

2/5    ▪ **1** ▪ 3 ▪ 5 ▪ 7 ▪ **9**

**35.** How many three-digit square numbers end in the digit 5?

▪ **2**

**36.** The square of a number is greater than or equal to the number.

1/3    ▪ Always true ▪ **Sometimes true** ▪ Never true

**37.** 1,225 is the product of two square numbers. What is the square root of 1,225?

▪ **35**

**38.** Think about what happens when you find the difference between the squares of two consecutive numbers.
What single calculation could you perform to calculate $97^2 - 96^2$?

▪ **96 + 97** ▪ **97 + 96**

**Level 5:**    Problem Solving - Involving the application of squares and square roots.

✸ **Required:** 5/5    ✸ **Student Navigation:** on
✸ **Randomised:** off

**39.** The pendant of a necklace is made by bending a piece of silver wire into three identical squares. If the total area inside the three squares is 507 mm² (square millimetres) how many millimetres of silver wire have been used?

▪ **156**

**40.** Ben starts working for his dad in April but his dad can't afford to give him his monthly wage all at once.
On the first day of the month he gives him 1p, on the second day 3p, on the third day 5p and so on. What single calculation can you perform to find out how much money he gets altogether?

▪ **30 × 30** ▪ **30 squared**

**41.** Two squares are placed together without overlapping to make a hexagon. The area of the hexagon is 85 cm². Find the **shortest** possible side of the hexagon in centimetres.

▪ **1**

**42.** A recipe for 6 portions of chocolate brownies requires a square-based tray with side length $s$. James wants to adapt the recipe to make 24 portions of brownies. What will the side-length of the new square-based tray need to be?

▪ **2 × s** ▪ **2s** ▪ **s × 2**

**Level 5:** *cont.*

**43.** Jodie has a square piece of card with an area of  171 cm². She want to cut out the largest possible square with an integer side. What is the length of the side in centimetres?

1/4

■ 13 ■ 14 ■ 85 ■ 86

# Understand the order of operations: brackets and powers

**Competency:**   Use convential notation for the priority of operations, including brackets, powers, roots and reciprocals

**Quick Search Ref:**   10084

**Correct:** Correct.   **Wrong:** Incorrect, try again.   **Open:** Thank you.

**Level 1:**   Understanding - Evaluating expressions with the four operations and brackets.

✿ **Required:** 7/10   ✿ **Student Navigation:** on   ✿ **Randomised:** off

---

**1.** In what **order** should you complete **operations** in an expression?

- Brackets ▪ Indices ▪ Division and Multiplication
- Addition and Subtraction

**2.** How could you write **3 × 3 × 3 × 3 × 3** using indices?

- $5^3$ ▪ $3^3$ ▪ $5^5$ ▪ $3^5$

1/4

**3.** To find the value of 17 + 21 × 13 - 9, what calculation would you do first?

- 17 + 21 ▪ 21 × 13 ▪ 13 - 9

1/3

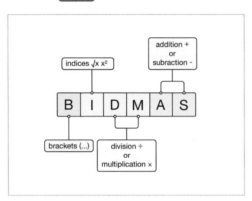

**4.** To find the value of 45 - 15 ÷ (3 + 2), what part of the expression would you calculate first?

- 45 - 15 ▪ 15 ÷ 3 ▪ 3 + 2

1/3

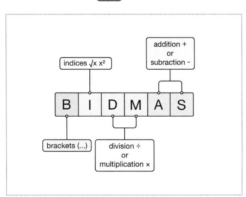

**5.** Calculate 11 + 7 × 5 - 2.

- 44

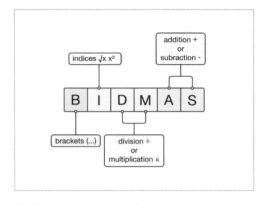

**6.** Calculate 72 - 48 ÷ (2 + 6).

- 66

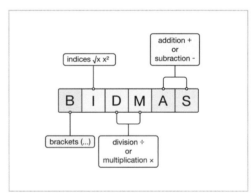

**7.** Find the value of 5 × (19 - 6) + 17.

- 82

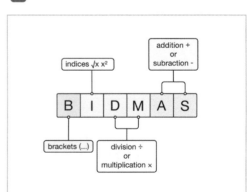

**8.** Calculate 81 - 54 ÷ 3 + 6.

- 69

---

**Level 1:** *cont.*

**9.** What is the value of  24 + 84 ÷ (12 - 8)?

 ▪ 45

**10.** Calculate 6 × 9 - 12 ÷ 2.

 ▪ 48

**Level 2:**  Fluency - Evaluating expressions that also include indices. decimals and operations within fractions.

✿ **Required:** 7/10    ✿ **Student Navigation:** on
✿ **Randomised:** off

**11.** What is the value of 2 × 3² - 13?

 ▪ 5

**12.** Calculate 4 × 3.7 + 2³.

 ▪ 22.8

**13.** Evaluate the expression shown.

 ▪ 21

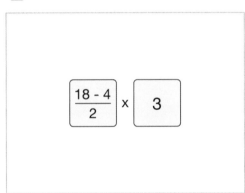

$$\frac{18 - 4}{2} \times 3$$

**14.** What is the value of 120 - (9 - 5)³ ÷ 8.

 ▪ 112

**15.** Which expression is equal to 30?

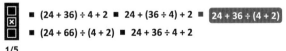 ▪ (24 + 36) ÷ 4 + 2  ▪ 24 + (36 ÷ 4) + 2  ▪ 24 + 36 ÷ (4 + 2)
▪ (24 + 66) ÷ (4 + 2)  ▪ 24 + 36 ÷ 4 + 2

1/5

**16.** Calculate 3 × (7 - 2.8) + 15.

 ▪ 27.6

**17.** Evaluate 9 × √36 ÷ 4.

 ▪ 13.5

**18.** What is the value of the expression shown?

 ▪ 10

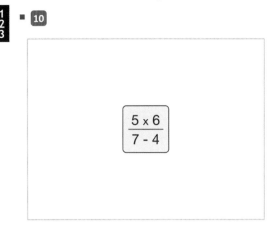

$$\frac{5 \times 6}{7 - 4}$$

**19.** Calculate 6² - 2.9 × 5.

 ▪ 21.5

**20.** Evaluate 27 + √144 ÷ 4 - 13.

▪ 17

**Level 3:**  Reasoning - Addressing misconceptions and evaluating multiple expressions.

✿ **Required:** 5/5    ✿ **Student Navigation:** on
✿ **Randomised:** off

**21.** Evaluate the following using the correct order of operations and then arrange your answers in **descending** order from top to bottom.

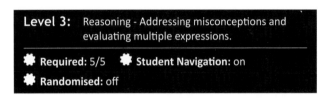 ▪ (9 + 4) × 6² - 13  ▪ (9 + 4) × (6² - 13)  ▪ 9 + 4 × 6² - 13
▪ 9 + 4 × (6² - 13)

**22.** Stuart says, "You can perform addition and subtraction in any order and get the same answer."
Is he correct? Explain your answer.

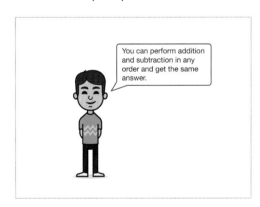

**23.** Which of the following statements is correct?

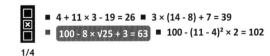 ▪ 4 + 11 × 3 - 19 = 26  ▪ 3 × (14 - 8) + 7 = 39
▪ 100 - 8 × √25 + 3 = 63  ▪ 100 - (11 - 4)² × 2 = 102

1/4

**Level 3:** *cont.*

**24.** Add one pair of brackets to the following expression to make it equal 28:

**a b c** $4 + 8 \times 5 - 2$

▪ 4 + 8 * (5 - 2)   ▪ 4 + 8 × (5 - 2)   ▪ 4 + 8 x (5 - 2)

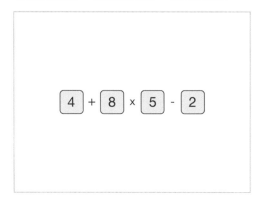

**25.** Karen says that $3^2 + 4^2 = (3 + 4)^2$.

**a b c** Is Karen correct? Explain your answer.

**Level 4:** Problem Solving - Multi-step problems involving the order of operations.

❋ **Required:** 5/5   ❋ **Student Navigation:** on
❋ **Randomised:** off

**26.** What missing number makes the statement below correct?

**1 2 3** $100 - ?^2 \div 4 + 7 \times 2 = 105$

▪ 6

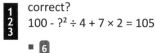

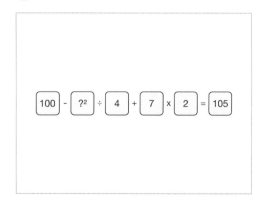

**27.** The answer to each calculation relates to a letter.

**a b c** Type the word the letters spell.

(i) $1.5 \times 2^3 - 7$

(ii) $9 + (12 - 4)^2 \div 16$

(iii) $8 \div \sqrt{4} + 9 \times 2$

(iv) $17 \times 3 - 5 \times 7$

(v) $2.4 \times 8 - 6 \div 5$

▪ match

| A | E | I | M | Q | U | Y |
|---|---|---|---|---|---|---|
| 13 | 23 | 17 | 5 | 31 | 11 | 20 |
| **B** | **F** | **J** | **N** | **R** | **V** | **Z** |
| 7 | 19 | 21 | 2 | 45 | 15 | 29 |
| **C** | **G** | **K** | **O** | **S** | **W** | |
| 16 | 4 | 9 | 49 | 25 | 2 | |
| **D** | **H** | **L** | **P** | **T** | **X** | |
| 14 | 18 | 3 | 57 | 22 | 67 | |

**28.** Add brackets to the following expressions to make them equal.

**1 2 3** What **value** does this give each expression?

$4 \times 25 + 40 \div 5 - \sqrt{9}$

$12 \div 2 + 4 \times 5^2 + 6$

▪ 130

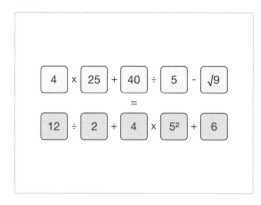

**29.** The numbers 1 to 6 can be represented by the following expressions.

↑ ↓ Evaluate each expression and then arrange them in **ascending order.**

▪ 44 / 44   ▪ (4 + 4 - 4) ÷ √4   ▪ (4 × 4 - 4) ÷ 4
▪ 4 + √4 × √4 - 4   ▪ (4 + 4 × 4) ÷ 4   ▪ (4 × 4 - 4) ÷ √4

**30.** Add operators to the following calculation to make it equal 60.

**a b c** Write the expression in full.

$9 \_\_ 8 \_\_ 4 \_\_ 3$

▪ 9 × 8 - 4 × 3   ▪ 9 x 8 - 4 x 3   ▪ 9 * 8 - 4 * 3

# Mathematics

**Y7**

## Algebra

Formulae

Expressions and Equations

Simplify

Sequences

Expand and Factorise

Solving Linear Equations

## Substitute Numerical Values (Advanced)

**Competency:**    Substitute numerical values into formulae and expressions, including scientific formulae.

**Quick Search Ref:**    10081

**Correct:** Correct.    **Wrong:** Incorrect, try again.    **Open:** Thank you.

**Level 1:**   Understanding - Understand how to substitute numerical values to evaluate formulas and expressions.

✸ **Required:** 7/10      ✸ **Student Navigation:** on      ✸ **Randomised:** off

---

**1.**   Give the value for the expression when:

1
2
3   $a = 10$ and $b = 2$.

▪ 5

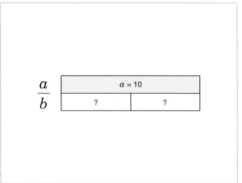

**2.**   Give the value for the expression when:

1
2
3   $s = 11$ and $t = 4$.

▪ 8.25

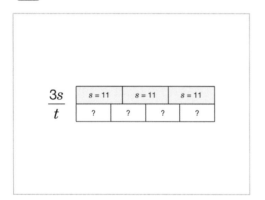

**3.**   Evaluate the expression when:

1
2
3   $s = 2$, $t = 4$, $u = 4$ and $v = 0.8$.

▪ 2.8

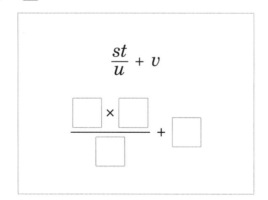

**4.**   Evaluate the expression when:

1
2
3   $a = 21$, $b = 5$ and $c = 3$.

▪ 12

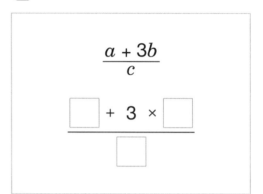

**5.**   Evaluate the expression when:

1
2
3   $a = 13$, $b = 7$, and $c = \frac{1}{2}$.

▪ 0.95

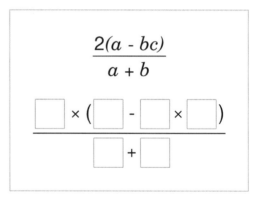

**6.**   Evaluate the expression when:

1
2
3   $n = 3$ and $m = 0.5$.

▪ 90

$$\frac{5n^2}{m}$$

$$\frac{5 \times \boxed{\phantom{x}}^2}{\boxed{\phantom{x}}}$$

**Level 1: *cont*.**

**7.** Evaluate the expression when:
$a$ = 8, $c$ = 40 and $d$ = 2.

```
1
2
3
```

▪ **7.5**

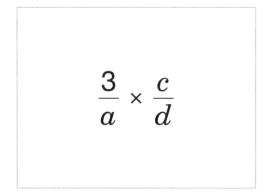

$$\frac{3}{a} \times \frac{c}{d}$$

**8.** Evaluate the expression when:
$d$ = 8, $e$ = 3 and $f$ = 4.
*Give your answer as a decimal.*

```
1
2
3
```

▪ **1.5**

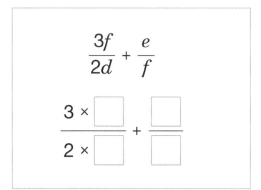

$$\frac{3f}{2d} + \frac{e}{f}$$

$$\frac{3 \times \square}{2 \times \square} + \frac{\square}{\square}$$

**9.** Evaluate the expression when:
$e$ = 3 and $f$ = ½.

```
1
2
3
```

▪ **100**

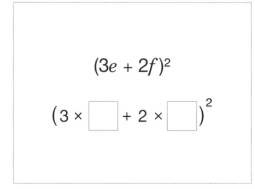

$$(3e + 2f)^2$$

$$\left(3 \times \square + 2 \times \square\right)^2$$

**10.** Evaluate the expression when:
$x$ = 2, $y$ = 5 and $z$ = 8.
*Give your answer as a decimal.*

```
1
2
3
```

▪ **-0.5**

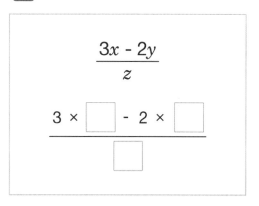

$$\frac{3x - 2y}{z}$$

$$\frac{3 \times \square - 2 \times \square}{\square}$$

**Level 2:** Fluency - Substitute numerical values including negatives and squares into more complex formulas and expressions.

✿ **Required: 7/10**   ✿ **Student Navigation:** on
✿ **Randomised:** off

**11.** Which of the cards gives a value of -25 if $n$ = -3?

```
☐
☒
☐
```

▪ A ▪ B ▪ C ▪ D ▪ **E** ▪ F ▪ G

1/7

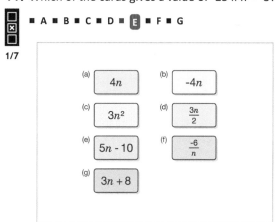

(a) $4n$
(b) $-4n$
(c) $3n^2$
(d) $\frac{3n}{2}$
(e) $5n - 10$
(f) $\frac{-6}{n}$
(g) $3n + 8$

**12.** Evaluate the expression when:
$a$ = 15 and $b$ = -2.

```
1
2
3
```

▪ **25**

$$\left(\frac{2a}{3b}\right)^2$$

$$\left(\frac{2 \times \square}{3 \times \square}\right)^2$$

**13.** Which of the cards is equal to 0.48 if $n = 0.4$?

■ A ■ B ■ **C** ■ D ■ E ■ F ■ G

1/7

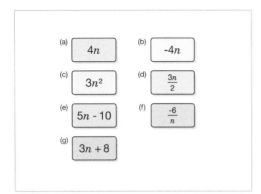

(a) $4n$

(b) $-4n$

(c) $3n^2$

(d) $\frac{3n}{2}$

(e) $5n - 10$

(f) $\frac{-6}{n}$

(g) $3n + 8$

**14.** Evaluate the expression when:
$m = -45$ and $n = -5$.

 ■ **4**

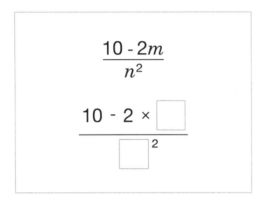

$$\frac{10 - 2m}{n^2}$$

$$\frac{10 - 2 \times \square}{\square^2}$$

**15.** Kinetic energy is given by the formula:
$k = \frac{1}{2}mv^2$. Calculate $k$ when $m = 8$ and $v = 5$.

 ■ **100**

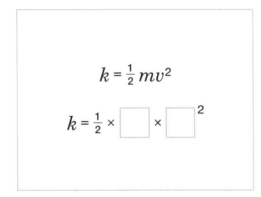

$$k = \frac{1}{2}mv^2$$

$$k = \frac{1}{2} \times \square \times \square^2$$

**16.** Evaluate acceleration using the formula:
$a = (v - u)/t$, when $v = 100$, $u = 22$ and $t = 5$.

 ■ **15.6**

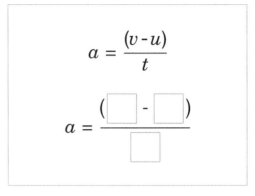

$$a = \frac{(v - u)}{t}$$

$$a = \frac{(\square - \square)}{\square}$$

**17.** Evaluate $s$ (the distance moved) if:
$u = 3.8$ m/s, $t = 2.5$ s and $a = 5.9$ m/s².
*Give your answer to two decimal places (2 d.p.).*

 ■ **27.94**

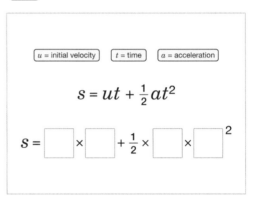

$u$ = initial velocity    $t$ = time    $a$ = acceleration

$$s = ut + \frac{1}{2}at^2$$

$$s = \square \times \square + \frac{1}{2} \times \square \times \square^2$$

**18.** The area (A) of a trapezium is given by the formula, where $a$ and $b$ are the lengths of the parallel sides and $h$ is the height of the trapezium. Find the area of trapezium T.
*Don't include the units in your answer.*

■ **24**

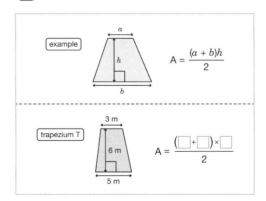

example

$$A = \frac{(a + b)h}{2}$$

trapezium T

3 m

6 m

$$A = \frac{(\square + \square) \times \square}{2}$$

5 m

**19.** Power is calculated using the formula:
$P = I^2R$, where $I$ is current and $R$ is resistance.
Evaluate $P$ when $I = 9$ and $R = ½$.

■ 40.5

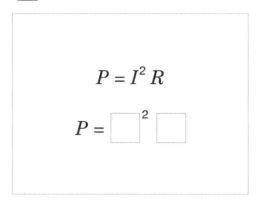

$$P = I^2 R$$

$$P = \Box^{2} \Box$$

**20.** The total surface area ($S$) of a cuboid is given by
the formula where $l$ is the length, $h$ is the height
and $w$ is the width.
Find the total surface area $S$ for cuboid $C$.
*Don't include the units in your answer.*

■ 174

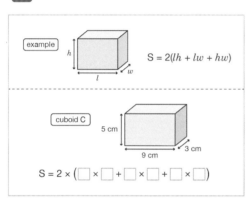

example
$$S = 2(lh + lw + hw)$$

cuboid C
5 cm
9 cm
3 cm

$$S = 2 \times (\Box \times \Box + \Box \times \Box + \Box \times \Box)$$

**Level 3:** Reasoning - Substitution into algebraic
fractions, checking accuracy and comparing
results.

✱ **Required:** 5/5   ✱ **Student Navigation:** on
✱ **Randomised:** off

**21.** Becky and Andy both have different answers to an
algebraic substitution problem. Who is correct?
Explain your answer.

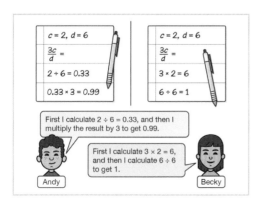

$c = 2, d = 6$
$$\frac{3c}{d} =$$
$2 ÷ 6 = 0.33$
$0.33 × 3 = 0.99$

$c = 2, d = 6$
$$\frac{3c}{d}$$
$3 × 2 = 6$
$6 ÷ 6 = 1$

First I calculate 2 ÷ 6 = 0.33, and then I
multiply the result by 3 to get 0.99.

Andy

First I calculate 3 × 2 = 6,
and then I calculate 6 ÷ 6
to get 1.

Becky

**22.** Sort the expressions in ascending order (smallest
value at the top).

↑
↓

■ D ▪ A ▪ C ▪ E ▪ B

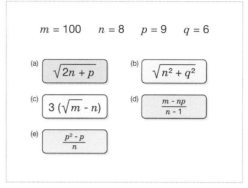

$m = 100$   $n = 8$   $p = 9$   $q = 6$

(a) $\sqrt{2n + p}$

(b) $\sqrt{n^2 + q^2}$

(c) $3(\sqrt{m} - n)$

(d) $\dfrac{m - np}{n - 1}$

(e) $\dfrac{p^2 - p}{n}$

**23.** What number is missing from the following
statement:
To make the expression true $n$ must must be a
positive number less than __?

■ 1

$m = 10$   $n = ?$

$$\frac{m}{n} > m$$

**24.** Which one of the following expressions is it not
possible to give a value for?

■ A ▪ B ▪ C ▪ D ▪ E ▪ F ▪ G

1/7

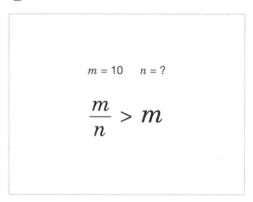

$a = -4$   $b = -25$   $c = \frac{1}{4}$

(a) $\sqrt{ab}$

(b) $\sqrt{c}$

(c) $-2b^2$

(d) $a^2$

(e) $\sqrt{a}$

(f) $3c^2$

(g) $\sqrt{abc}$

**25.** Which expression is the odd one out? Explain your
answer.

a
b
c

**Reminder**: √100 is the number that when
multiplied by itself gives 100.

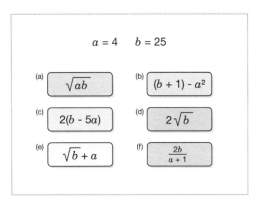

$a = 4$    $b = 25$

(a) $\sqrt{ab}$

(b) $(b + 1) - a^2$

(c) $2(b - 5a)$

(d) $2\sqrt{b}$

(e) $\sqrt{b} + a$

(f) $\frac{2b}{a + 1}$

**Level 4:** Problem Solving - Substitution into algebraic
formulae in context and taking further steps
to solve a problem.

✱ **Required:** 5/5  ✱ **Student Navigation:** on
✱ **Randomised:** off

**26.** The longest side of a right-angled triangle can be
found by using the formula:

1
2
3

$c = \sqrt{(a^2 + b^2)}$

Find the perimeter of square A, when $a = 3$ cm and
$b = 4$ cm.

*Don't include the units in your answer.*

▪ 20

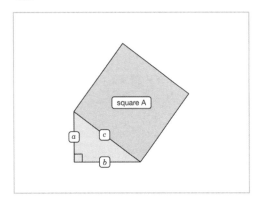

square A

a    c

b

**27.** Nadia is planting seeds which will germinate at
temperatures above 50°F, but her thermometer
only has a centigrade scale and also gives
inaccurate readings, as much as 3°C above the
actual temperature. When taking a reading from
her thermometer, at what temperature can Nadia
be confident that the seeds will germinate?

1
2
3

▪ 13

To convert degrees fahrenheit to
degrees celsius, use the formula:

$$C° = \frac{5(f° - 32)}{9}$$

**28.** The area of trapezium B is equal to the area of
rectangle C. Find the length $x$ for a side of
rectangle C.

1
2
3

*Don't include the units in your answer.*

▪ 3

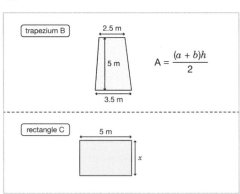

trapezium B

2.5 m

5 m

3.5 m

$A = \frac{(a + b)h}{2}$

rectangle C

5 m

x

**29.** The volume of cuboid X is equal to the volume of
cube Y.

1
2
3

Find the length $y$ for a side of cube Y when $l = 9$
cm, $w = 4$ cm and $h = 6$ cm.

*Don't include the units in your answer.*

▪ 6

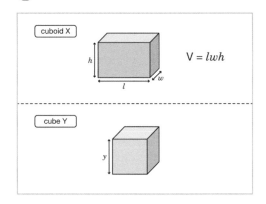

cuboid X

h

l

w

$V = lwh$

cube Y

y

**30.** Charlie weighs 187 lbs and is 6 feet 2 inches tall (6' 2"). What is his BMI in kg/m²?
*Round your answer to 2 decimal places (2 d.p.) and don't include the units in your answer.*
**Hint:** You need to convert to metric units before you can use the formula.

■ 24.06

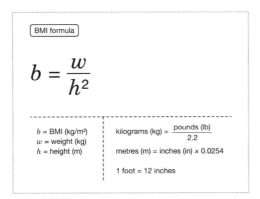

# Substitute Numerical Values (Basic)

**Competency:**   Substitute numerical values into formulae and expressions, including scientific formulae.

**Quick Search Ref:**   10243

**Correct:** Correct.     **Wrong:** Incorrect, try again.     **Open:** Thank you.

**Level 1:**   Understanding - Substitute numerical values into formulas and expressions.

⚙ **Required:** 7/10          ⚙ **Student Navigation:** on          ⚙ **Randomised:** off

---

**1.**   If $x = 8$, what is the value of $6x$?

 ▪ 48

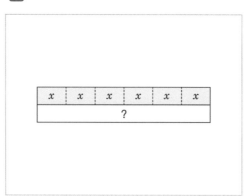

**2.**   If $x = 9$, evaluate $2x + 7$.

 ▪ 25

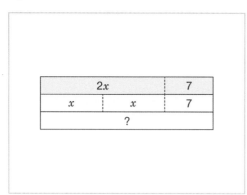

**3.**   If $n = 100$, what is the value of $n - 6$?

 ▪ 94

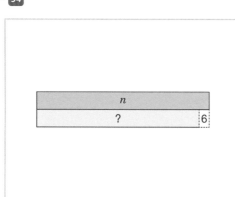

**4.**   If $w = 8$, evaluate $9 - w$.

 ▪ 1

**5.**   If $y = 16$, evaluate $y/2$.

 ▪ 8

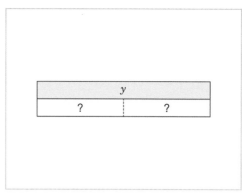

**6.**   If $a = 7$, what is the value of $3 + 2a$?

 ▪ 12  ▪ 35  ▪ 17  ▪ 30

1/4

**7.** If $x = 11$, evaluate $3(x + 5)$.

 ▪ 48

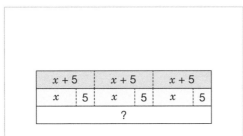

**8.** If $x = 3$, evaluate $27 - 2x$.

 ▪ 21

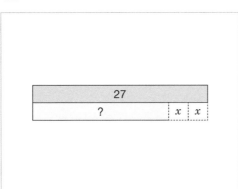

**9.** If $y = 7$, evaluate $y^2$.

 ▪ 49

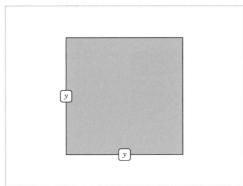

**10.** If $f = 5$, evaluate $4f^2$.

 ▪ 100

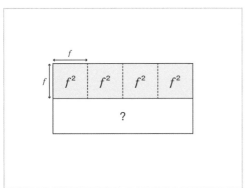

**Level 2:**    Fluency - Substitution of negatives, decimals and fractions. Expressions with squared variables and brackets.

✱ **Required:** 7/10    ✱ **Student Navigation:** on
✱ **Randomised:** off

**11.** If $x = -8$, what is the value of $6x$?

 ▪ -48

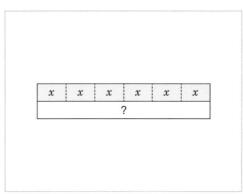

**12.** If $y = 0.75$, what is the value of $2(y + 3)$?
Write your answer as a decimal.

 ▪ 7.5

**13.** If $t = 46$, what is the value of $\frac{1}{2}t - 6$?

 ▪ 17

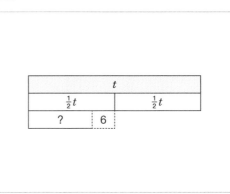

**14.** If $x = 4$, evaluate $5(9 - 3x)$.

 ▪ -15

**15.** If $d = -¼$, evaluate $3d - 4$.

Give your answer as a decimal.

 ▪ -4.75

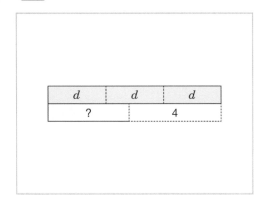

**16.** If $x = -4$, evaluate $x^2$.

▪ 16

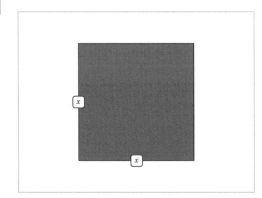

**17.** Calculate the cost of $5p + 2r$.

$p$ = pencil
$r$ = ruler

*Give your answer in pounds and pence. Don't forget to include the £ sign in your answer.*

▪ £7.91

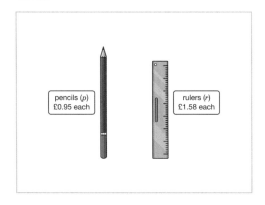

**18.** Calculate the cost of $4w + c$.

$c$ = cake
$l$ = lemonade
$p$ = pineapple
$w$ = water

*Give your answer in pounds and pence. Don't forget to include the £ sign in your answer.*

▪ £7.15

**19.** Calculate the cost of $3p + 2l + w$.

$c$ = cake
$l$ = lemonade
$p$ = pineapple
$w$ = water

*Give your answer in pounds and pence. Don't forget to include the £ sign in your answer.*

▪ £14.85

**Level 2: *cont.***

**20.** Calculate the cost of 3(p + w).

c = cake
l = lemonade
p = pineapple
w = water

*Give your answer in pounds and pence. Don't forget to include the £ sign in your answer.*

■ **£8.85**

**Level 3:** Reasoning - Substitution in context.

✱ **Required: 5/5**   ✱ **Student Navigation: on**

✱ **Randomised: off**

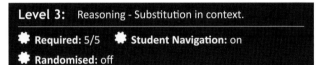

**21.** Nadia and Tom go to the cinema. Nadia pays for the tickets: 2(c + b). Tom pays for the food: 4l + 2p + h + i.

How much more does Nadia spend than Tom?
*Give your answer in pounds and pence. Don't forget to include the £ sign in your answer.*

■ **£3.20**

**22.** The area of the piece of paper is given as:
xy = 16 (x and y are both integers).
Find the two value sets for x and y that give an area of 16.

2/7

■ x = 8, y = 8  ■ x = 0, y = 16  ■ x = 1, y = 15  ■ **x = 1, y = 16**
■ x = 2, y = 14  ■ **x = 4, y = 4**  ■ x = 3, y = 13

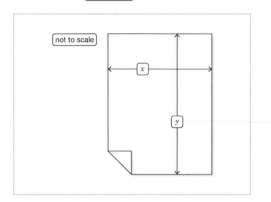

**23.** Khalid and Stefan are discussing an algebra problem. Who is incorrect? Explain your reasoning.

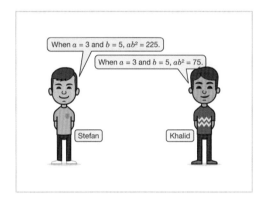

**24.** Select the options where $ab^2 = (ab)^2$.

■ a = 2 and b = 2  ■ **a = 1 and b = 5**  ■ a = 2 and b = 1
■ a = -1 and b = -1  ■ a = ½ and b = ½  ■ **a = 5 and b = 0**

2/7  ■ a = -½ and b = ½

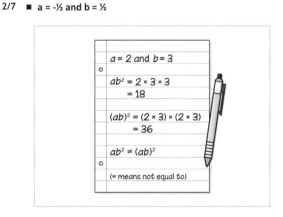

Level 3: *cont.*

**25.** Tara and Paula are discussing how to write a formula for working out the number of months for any given number of years.
Explain why Tara is incorrect. What should the formula be?

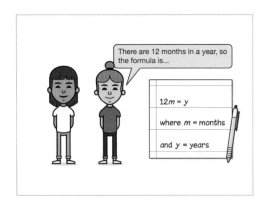

**Level 4:** Problem Solving - Substitution into formulae.

✿ **Required:** 5/5    ✿ **Student Navigation:** on
✿ **Randomised:** off

**26.** The area of any triangle is given by the formula Area = ½bh where *h* stands for the perpendicular height and *b* stands for the length of the base. Calculate the area of triangle B in square centimetres (cm²).
*Don't include the units in your answer.*

▪ 35

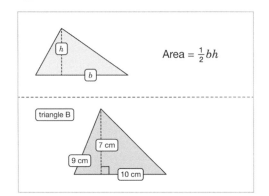

**27.** The perimeter of a rectangle is given by the formula *perimeter = 2l + 2w* where *l* stands for length and *w* stands for width. Calculate the perimeter of rectangle A.
*Include the units cm (centimetres) in your answer.*

▪ 48 cm

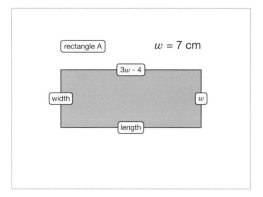

**28.** The illustration shows that if you want to send a parcel to an address 9 miles away, Deliver-Fast are cheaper than Speedy Parcel. After how many miles will Speedy Parcel be cheaper? Give reasons for your answer.

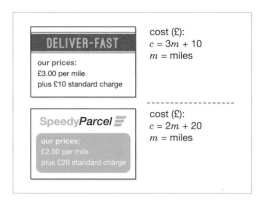

**29.** Gary is travelling to see an old friend. Every time he stops for a break he notes the distance travelled (in miles) and the time taken for that part of the journey. What is Gary's average speed for the whole journey in miles per hour (mph)?
*Include the units mph (miles per hour) in your answer.*

▪ 35 miles per hour  ▪ 35 mph

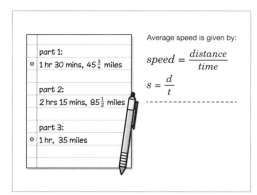

**30.** The area of trapezium C is equal to the area of square D. Find the value *x*, which is the length of a side of square D.

*Don't include the units in your answer.*

▪ **6**

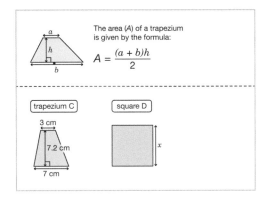

The area (A) of a trapezium is given by the formula:

$$A = \frac{(a + b)h}{2}$$

trapezium C

3 cm

7.2 cm

7 cm

square D

*x*

# Use Simple Formulae

**Competency:**  Use simple formulae.

**Quick Search Ref:**  10228

**Correct:** That's right.  **Wrong:** No. Try again?  **Open:** Thank you.

**Level 1:**  Understanding - Solve equations with one letter.

✸ **Required:** 9/12  ✸ **Student Navigation:** on  ✸ **Randomised:** off

**1.** In the diagram for 5*a*, *a* is _____ 5.

■ added to ■ divided by ■ multiplied by
■ subtracted from

1/4

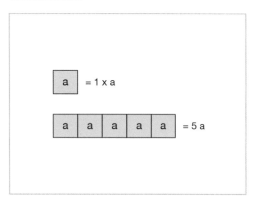

**2.** In the diagram *x*/3, *x* is _____ 3.

■ added to ■ divided by ■ multiplied by
■ subtracted from

1/4

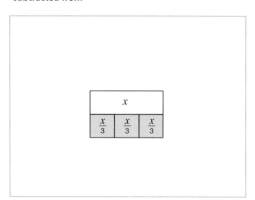

**3.** What is the value of *b* if *b* - 2 = 6?

■ 3 ■ 4 ■ 8 ■ 12

1/4

**4.** What is the value of *y* if *y* + 3 = 9?

■ 6

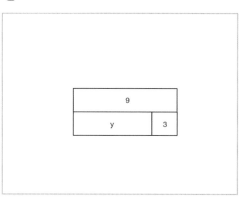

**5.** What is the value of *a* if 4*a* = 12?

■ 3 ■ 8 ■ 16 ■ 48

1/4

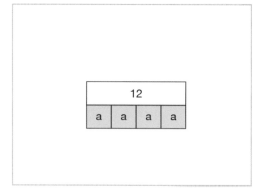

**6.** 7*y* = 35.

What is the value of *y*?

■ 5

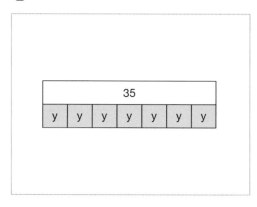

**7.** What is the value of *a* if *a*/5 = 10?

■ 2 ■ 5 ■ 15 ■ 50

1/4

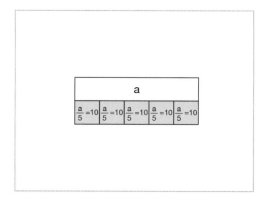

**8.** What is the value of *x* if *x*/3 = 12?

 ▪ 36

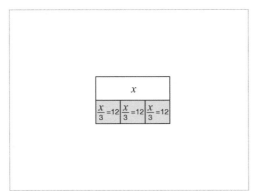

**9.** What is **5a** if *a* = 8?

 ▪ 40

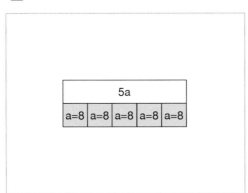

**10.** 6*y* = 24.

 What is the value of *y*?

▪ 4

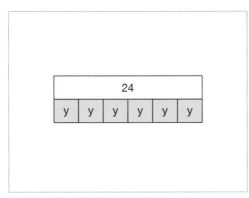

**11.** *b*/4 = 4.

 What is the value of *b*?

▪ 16

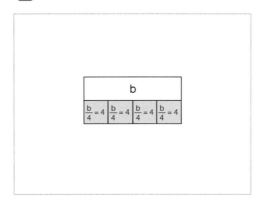

**12.** What is **4x** if *x* = 12?

 ▪ 48

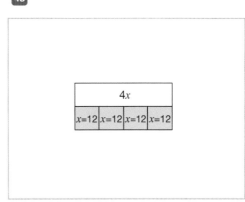

**Level 2:** Fluency - Solve equations with one letter plus another function.

❋ **Required:** 8/11    ❋ **Student Navigation:** on
❋ **Randomised:** off

**13.** If *x* = 6, what is **x - 2?**

 ▪ 4

**14.** If 4*a* + 4 = 20, what does *a* equal?

 ▪ 4

**15.** When a number goes in the number machine it **subtracts 5** from the number.

The equation for this machine is *a* - 5 = *b*.

The number 17 is put into the machine. If *a* = 17, what is the value of *b*?

▪ 12

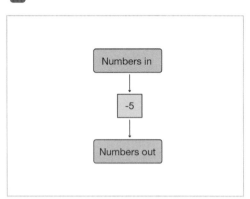

**16.** What is the value of **x** when 7*x* + 28 = 70?

▪ 6

**17.** When a number goes in the number machine it **adds 12** to the number.

The equation for this machine is *c* + 12 = *d*.

The number 14 is put into the machine. If *c* = 14, what is the value of *d*?

▪ 26

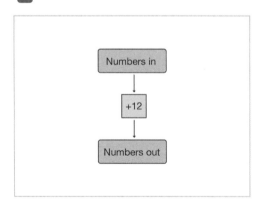

**18.** If **y** = 8, what is *y* + 13?

▪ 21

**19.** Find the value of **y** in the following equation:

80 = 8*y* +16.

▪ 8

**20.** When a number goes in the number machine it **adds 12** to the number.

The equation for this machine is *e* + 12 = *f*.

The number 32 comes out of the machine.

Calculate the value of *e* when *f* = **32** to find the number that went into the machine.

▪ 20

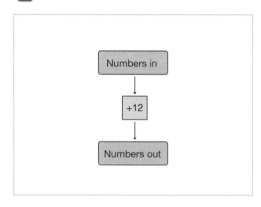

**21.** What is **z** - **15** when *z* = 24?

▪ 9

**22.** If 5*a* - 5 = 30, what does **a** equal?

▪ 7

**23.** When a number goes in the number machine it **subtracts 5** from the number.

The equation for this machine is *a* - 5 = *b*.

The number 20 comes out of the machine.

Calculate the value of *a* when *b* = **20** to find the number that went into the machine.

▪ 25

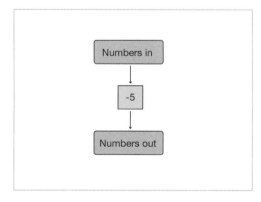

**24.** An apple costs *x* amount.

1/4

**Three apples cost 75 pence**. Select the correct equation which represents this problem.

■ x = 3  ■ x = 25  ■ x = 75  ■ 3x = 75

**25.** Calculate *x* for each equation.

Order the values of *x* from smallest to largest.

■ 5x + 1 = 6  ■ 3x + 7 = 16  ■ 2x - 5 = 3

**26.** $2b = b^2$.

Is this statement correct? Explain your answer.

**27.** *a* is the number of spectators who are allowed into a concert. There are also 28 staff so the formula for people at the concert is:

*a* + 28.

If the maximum capacity of the venue is 216 people, what is the **maximum** number of spectators allowed inside the venue?

■ 188

**28.** Calculate *y* for each equation.

**Order** the values of *y* from smallest to largest.

■ 4y + 7 = 19  ■ 3y - 8 = 10  ■ 5y + 50 = 100

**29.** Class 6 are split into 4 **equal** groups represented by the letter *y*.

There is one teacher and three teaching assistants in the class so the formula for people in the class is:

**4y + 4.**

How many children are in each group if there are **24 people** in the class?

■ 5

**30.** A quadrilateral has two equal sides each measuring *a* and two equal sides each measuring *b*.

1/4

Which of these formulae would you use to calculate its **perimeter**?

■ 2ab  ■ ab  ■ 2a + 2b  ■ a + b

**31.** A bottle of water costs *x* amount.

1/4

5 bottles of water cost £2.00. Select the correct equation which represents this problem.

■ x = 5  ■ x = 40p  ■ x = 2  ■ 5x = 2

**32.** Sarah says "If $3b + b = 32$ and $3b = 24$, the value of *b* has to be 8."

Is she correct? Explain your answer.

**33.** Each shape in the grid has a value. The **sum** of their values are at the end of some rows and columns.

Use these totals to find the value of a **square**.

■ 11

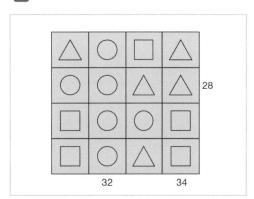

**34.** Using the equation $4a + 4 = 40$, calculate the value of *b* in the following equation:

$ab + 7 = 70$.

■ 7

**35.** Each symbol in the grid has a value. The sum of their values are at the end of some rows and columns.

Use these totals and find the value of a **flag**.

■ 15

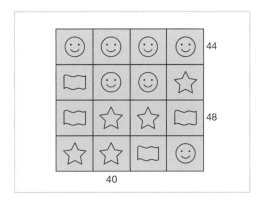

**36.** Using the equation $5a + 7 = 62$, calculate the value

of **b** in the following equation:

$ab - 22 = 55$.

▪ 7

**37.** Frank has a fruit and vegetable stall and makes

some packs to sell. Look at the image.

Using the information from the image, find the
cost of **1 peach**. Remember to include the unit
pence (p) in your answer.

▪ 42pence ▪ 42p

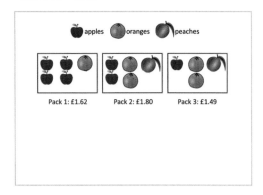

**38.** Look at the image of a breakfast order.
Using the information, complete the sentence
below.

1 portion of **milk** costs ___ pence.

▪ 12

| ORDER: 321 | | |
|---|---|---|
| Sam: | 4 x toast; 1 x egg | COST: £1.32 |
| Arnold: | 3 x toast; 1 x milk; 2 x eggs | COST: £1.51 |
| Fiona: | 1 x milk; 2 x eggs; 2 x toast | COST: £1.26 |

# Form Expressions

**Competency:** Know the difference between variable, term, coefficient and expression; understand basic algebraic notation and form expressions from situations described in words.

**Quick Search Ref:** 10021

**Correct:** Correct.    **Wrong:** Incorrect, try again.    **Open:** Thank you.

**Level 1:** Understanding - Form expressions from mathematical vocabulary.

✿ **Required:** 7/10          ✿ **Student Navigation:** on          ✿ **Randomised:** off

1.  Which expression means **d add 2**?

    ■ d + 2 ■ 2d ■ d² ■ d - 2

    1/4

    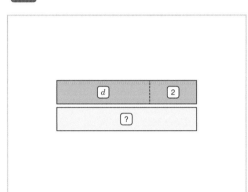

2.  Select an algebraic expression for **6 subtracted from m**.

    ■ 6 - m ■ m + 6 ■ -6m ■ m - 6
    1/4

    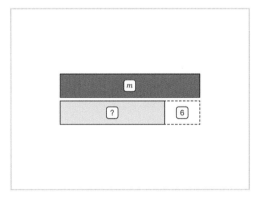

3.  Which term best describes **4 multiplied by z**?

    ■ z4 ■ z⁴ ■ 4z ■ zzzz
    1/4

    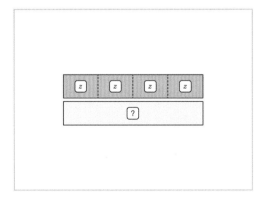

4.  What is an expression for **c divided by 3**?

     *To type a divide sign, use the / key.*
    ■ c/3

    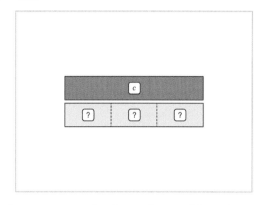

5.  Form an expression for **s subtracted from 9**.

     ■ 9 - s

    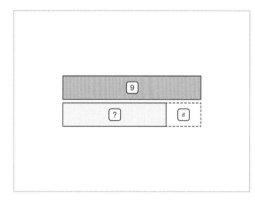

6.  What expression is the same as **4 more than a**?

     ■ a + 4 ■ 4 + a

7.  What is an expression for **5 divided by t**?

     *To type a divide sign, use the / key.*
    ■ 5/t

8.  What is an expression for **6 less than r**?

     ■ r - 6

9.  Form an expression for **v shared between 5**.

     ■ v/5

**10.** Give an expression for *k less than 7*

a
b
c
■ `7 - k`

**Level 2:** Fluency - Form expressions from worded situations.

✱ **Required:** 7/10  ✱ **Student Navigation:** on
✱ **Randomised:** off

**11.** Form an expression for the following:
8 lots of *h* minus 5.
a
b
c
■ `8h - 5`  ■ `-5 + 8h`

**12.** Eva is *x* years old and her brother Harrison is 3 years older than her.
☐
☒
☐
What expression represents Harrison's age?
1/4  ■ `3`  ■ `x - 3`  ■ `3x`  ■ `x + 3`

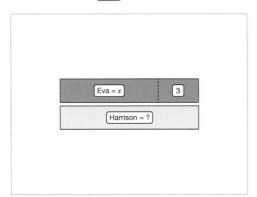

**13.** A football coach has *b* footballs but she loses 2 of them.
☐
☒
☐
Select the expression that represents how many balls she has left.
1/4  ■ `2 - b`  ■ `2b`  ■ `b - 2`  ■ `b + 2`

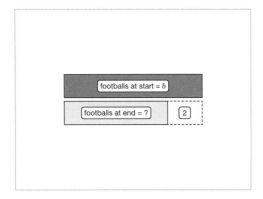

**14.** A square has side length *k*.
a
b
c
What is an expression for the perimeter of the square?
*Answer in the simplest form.*

■ `4k`

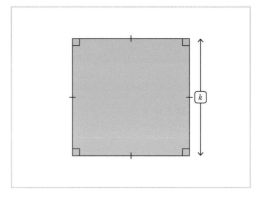

**15.** Roman is twice the age of Annabel and Theo is 3 years older than Roman.
a
b
c
If Annabel is *y* years old, what is an expression for Theo's age?

■ `3 + 2y`  ■ `2y + 3`

**16.** Form an expression for the perimeter of the triangle.
a
b
c
*Answer in the simplest form.*

■ `n + 17`  ■ `17 + n`

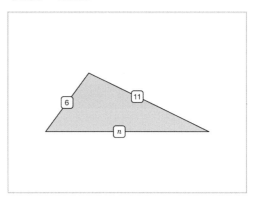

**17.** Aisha is 5 centimetres (cm) taller than Isabel. If Aisha has a height of *h* cm, give an expression for the height of Isabel.
a
b
c

■ `h - 5`  ■ `h - 5 cm`  ■ `h - 5 centimetres`

**Level 2:** *cont.*

**18.** What is an expression for the perimeter of the
a
b triangle?
c *Answer in the simplest form.*

- 2a + 5 ▪ 5 + 2a

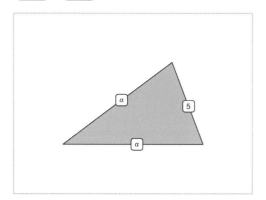

**19.** A square has a side length of *z*.
Which expression gives the area of the square?

- z² ▪ 4z ▪ z × z

1/3

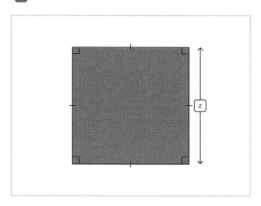

**20.** One side of an equilateral triangle is *r* + 4.
a Form an expression for the perimeter of the
b triangle.
c

- 3r + 12 ▪ 12 + 3r

**Level 3:** Reasoning - Forming expressions with two
operators.

❁ **Required:** 5/5    ❁ **Student Navigation:** on
❁ **Randomised:** off

**21.** A plumber has a call-out charge of £30 and then
charges £20 per hour to complete repairs.
Which expression represents the total cost of a
repair taking *h* hours?

1/4

- 50 ▪ 50h ▪ 30h + 20 ▪ 20h + 30

**22.** The area of the square is *a* and the area of the
a quadrant is *b*.
b
c Form an expression for the area of the striped
shape.

- a - b

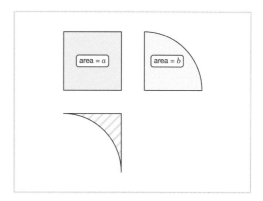

**23.** Javier has written an expression for the output of
a the function machine.
b
c Is his expression correct? Explain your answer.

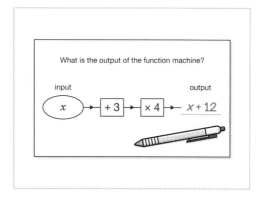

**24.** Jack is thinking of two consecutive numbers.
a The largest number is *s*.
b
c Give an expression for the sum of the two
numbers.

- 2s -1 ▪ 2s + 1

**25.** A function machine has two stages:
a 1. Subtract 7.
b 2. Multiply by 3.
c If *n* is the input, what is the output?

- 3n - 21 ▪ -21 + 3n ▪ 3 x (n - 7) ▪ 3(n - 7)
- 3 × (n - 7) ▪ 3 * (n - 7)

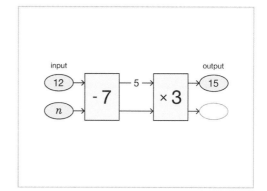

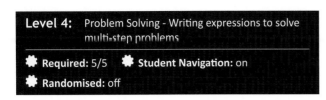

**Level 4:** Problem Solving - Writing expressions to solve multi-step problems

✹ **Required:** 5/5    ✹ **Student Navigation:** on
✹ **Randomised:** off

**26.** What is an expression for the area of the coloured part of the shape?

a
b
c

▪ 3w

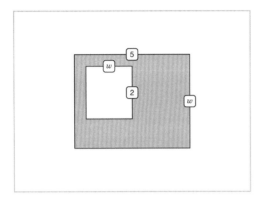

**27.** Give an expression for the size of angle *t* in degrees.

a
b
c

*Don't include the units in the expression.*

▪ 180 - 2s

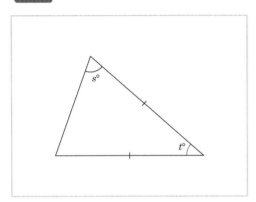

**28.** The diagram shows a rectangle of length $x + 9$ cm and height $x$ cm. If the perimeter of the rectangle is 42 cm, what is the value of $x$?

1
2
3

▪ 6

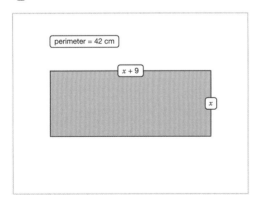

**29.** Bilal has twice as many merits as Archie, and Cody has three times as many merits as Bilal. If the three boys have 108 merits between them, how many merits does Cody have?

1
2
3

▪ 72

**30.** Two rectangles are joined to make the shape shown. If the perimeter of the shape is 50 cm, what is the value of *a*?

a
b
c

*Include the units cm (centimetres) in your answer.*

▪ 7 cm   ▪ 7 centimetres

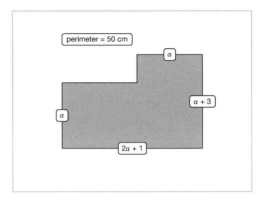

# Find the nth Term of Linear Sequences

| **Competency:** | Recognise arithmetic sequences and find the nth term. |
| --- | --- |

| **Quick Search Ref:** | 10255 |
| --- | --- |

Correct: Correct.　　Wrong: Incorrect, try again.　　Open: Thank you.

**Level 1:**　Understanding - Finding the nth term in basic number sequences.

✹ **Required:** 7/10　　✹ **Student Navigation:** on　　✹ **Randomised:** off

---

**1.** What is the common difference in the following arithmetic sequence?

1
2
3

3, 6, 9, 12, 15, ...

▪ **3**

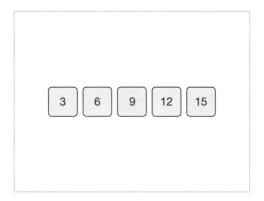

**2.** What is the 100th term in the following arithmetic sequence?

1
2
3

3, 6, 9, 12, 15, ...

▪ **300**

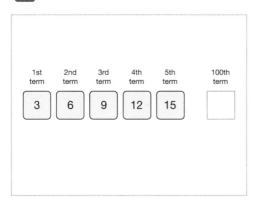

**3.** What is the *n*th term in the following arithmetic sequence?

a
b
c

3, 6, 9, 12, 15, ...

▪ **3n**

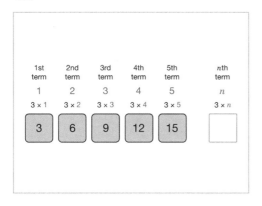

**4.** What is the *n*th term in the following arithmetic sequence?

a
b
c

4, 7, 10, 13, 16 ...

▪ **3n + 1**

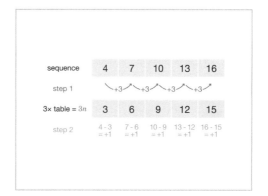

**5.** What is the *n*th term in the following arithmetic sequence?

a
b
c

7, 11, 15, 19, 23, ...

▪ **4n + 3**

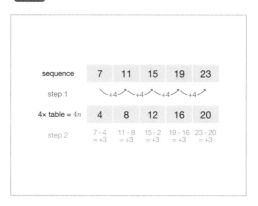

**6.** What is the *n*th term in the following arithmetic sequence?

a
b
c

8, 15, 22, 29, 36, ...

▪ **7n + 1**

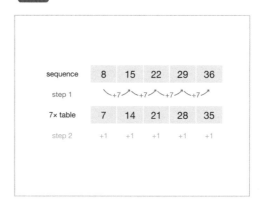

---

**7.** What is the *n*th term in the following arithmetic
sequence?
**a b c**
2, 5, 8, 11, 14, ...

■ 3n - 1

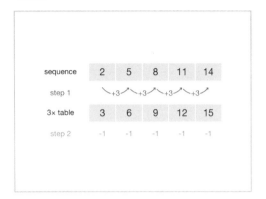

**8.** What is the *n*th term in the following arithmetic
sequence?
**a b c**
11, 17, 23, 29, 35, ...

■ 6n + 5

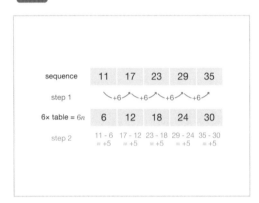

**9.** What is the *n*th term in the following arithmetic
sequence?
**a b c**
5, 7, 9, 11, 13, ...

■ 2n + 3

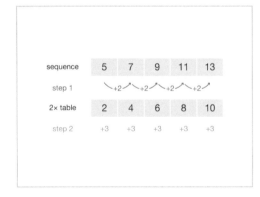

**10.** What is the *n*th term in the following arithmetic
sequence?
**a b c**
3, 8, 13, 18, 23, ...

■ 5n - 2

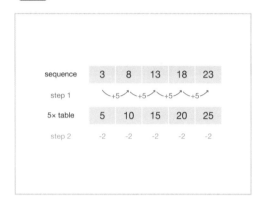

**Level 2:**  Fluency - Finding the nth term in more
complex sequences and in sequences of
patterns.

✿ **Required:** 7/10    ✿ **Student Navigation:** on
✿ **Randomised:** off

**11.** What Is the *n*th term in the following arithmetic
sequence?
**a b c**
9, 13, 17, 21, 25, ...

■ 4n + 5

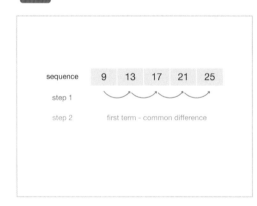

**12.** What is the *n*th term in the following arithmetic
sequence?
**a b c**
7, 16, 25, 34, 43, ...

■ 9n - 2

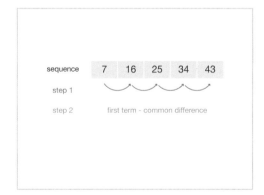

**13.** What is the *n*th term in the following arithmetic
**a** sequence?
**b**
**c** 6, 7, 8, 9, 10, …

■ `1n + 5`  ■ `5 + n`  ■ `n + 5`

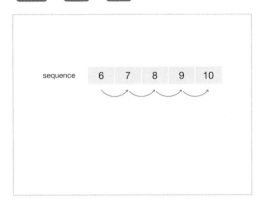

**14.** How many matchsticks would you need to make
**a** the *n*th term in the sequence shown?
**b**
**c** ■ `2n + 1`

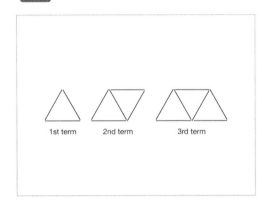

**15.** What is the *n*th term in the following arithmetic
**a** sequence?
**b**
**c** 14, 11, 8, 5, 2, …

■ `17 - 3n`  ■ `-3n + 17`

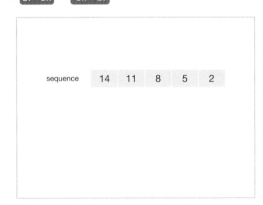

**16.** What is the *n*th term in the following arithmetic
**a** sequence?
**b**
**c** 3, 2, 1, 0, -1, …

■ `4 - n`  ■ `-1n + 4`  ■ `4 - 1n`  ■ `-n + 4`

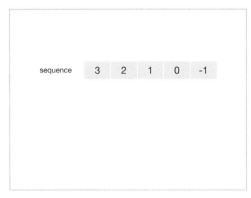

**17.** The diagram shows how many chairs can be
**a** placed around tables arranged in rows. How many
**b** people can be seated at a row of *n* tables?
**c**
■ `2 + 2n`  ■ `2n + 2`

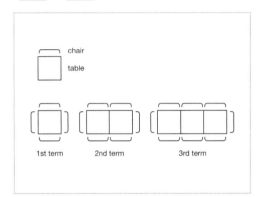

**18.** What is the *n*th term in the following sequence?
**a** 2, -3, -8, -13, -18, …
**b**
**c** ■ `7 - 5n`  ■ `-5n + 7`

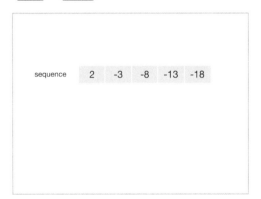

Level 2: *cont*.

**19.** How many dots make the *n*th term in this sequence?

■ 4n + 1   ■ 1 + 4n

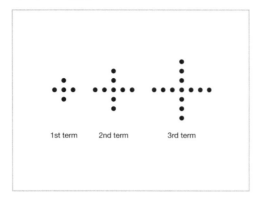

1st term        2nd term        3rd term

**20.** What is the *n*th term in the following sequence?
-3, -4, -5, -6, -7, ...

■ -1n - 2   ■ -2 - 1n   ■ -2 - n   ■ -n - 2

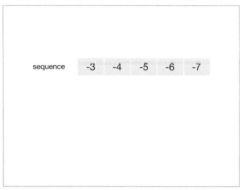

sequence    -3    -4    -5    -6    -7

**Level 3:**  Reasoning - Misconceptions and reasoning with sequences.

✿ **Required:** 5/5   ✿ **Student Navigation:** on
✿ **Randomised:** off

**21.** Noah says the sequence of square numbers is not an arithmetic sequence.

Is Noah correct? Explain your answer.

**22.** Find the 100th term in the following sequence:
-3.25, -1.75, -0.25, 1.25, 2.75, ...
*Give your answer as a decimal.*

■ 145.25

sequence    -3.25  -1.75  -0.25  1.25  2.75

**23.** Jayne generates the sequence 4*n* + 3 as follows:
7, 11, 15, 19, 23, ...

She says, "The 5th term in the sequence is 23, so the 50th term will be 230".

Is Jayne correct? Explain your answer.

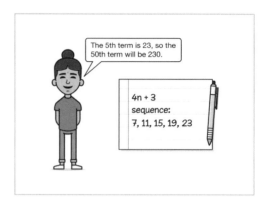

**24.** The first term in an arithmetic sequence is 7 and the fourth term is 19.

What is the *n*th term in the sequence?

■ 4n + 3

**25.** Sort the sequences in ascending order (smallest first) according to the value of the 25th terms.

■ 79, 78, 77, 76, 75, ...   ■ 8, 10, 12, 14, 16

■ -15, -12, -9, -6, -3, ...

**Level 4:**  Problem Solving - Multi-step problems with sequences.

✿ **Required:** 5/5   ✿ **Student Navigation:** on
✿ **Randomised:** off

**26.** Sian is looking at the following sequence:
6, 13, 20, 27, 34 ...
What position in the sequence is the first term with a value over 500?

■ 72

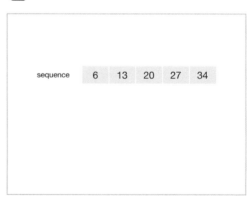

sequence    6    13    20    27    34

**27.** The fifth term in an arithmetic sequence is 19 and the tenth term is 39.

What is the *n*th term in the sequence?

■ 4n - 1

**Level 4:** *cont.*

**28.** Here are the first 5 terms of three arithmetic
a
b
c
sequences mixed together.

1, 1, 3, 4, 4, 5, 6, 7, 7, 8, 9, 10, 10, 12, 13

What is the *n*th term in the sequence with the
greatest common difference?

▪ `3n - 2`  ▪ `-2 + 3n`

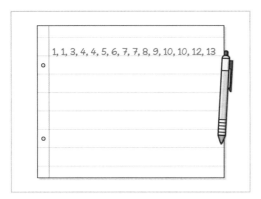

**29.** The first five terms in an arithmetic sequence have
a
b
c
a common difference of +3 and a sum of 20. What
is the formula for the *n*th term in the sequence?

▪ `3n - 5`  ▪ `-5 + 3n`

**30.** The sequences 3*n* - 2 and 2*n* + 5 have some terms
a
b
c
in common.

What is the *n*th term for the series of numbers the
sequences have in common?

▪ `6n + 1`  ▪ `1 + 6n`

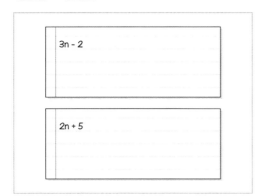

# Generate Sequences From the nth Term

**Competency:**    Generate terms of a sequence from either a term-to-term or a position-to-term rule.

**Quick Search Ref:**    10110

Correct: Correct.     Wrong: Incorrect, try again.     Open: Thank you.

**Level 1:**    Understanding - Using the nth term rule to find missing terms in a simple sequence.

✿ **Required:** 7/10       ✿ **Student Navigation:** on       ✿ **Randomised:** off

---

**1.** Sort the following words to correctly match the order of definitions shown in the diagram.

↑
↓

- Arithmetic sequence ▪ Term ▪ Common difference
- Substitution

| |
|---|
| A pattern of numbers which increase or decrease by the same amount each time. |
| A word to describe each number in a sequence. |
| The difference between each term in an arithmetic sequence. |
| To replace a letter with a number when using algebra. |

**2.** Which one of the following is **not** an arithmetic sequence?

☐
☒
☐

1/6

- 2, 5, 8, 11 ▪ 4, 8, 12, 16 ▪ 1, 2, 3, 4 ▪ 3, 6, 9, 11
- 10, 8, 6, 4 ▪ -1, 1, 3, 5

**3.** If $n = 3$, what is the value of $2n + 4$?

☐
☒
☐

1/5

- 14 ▪ 9 ▪ 10 ▪ 27 ▪ 6

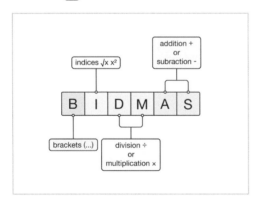

**4.** The table shows the first four terms of the arithmetic sequence $3n - 1$.
Select the statement that is **not** true.

☐
☒
☐

1/4

- The 1st term in the sequence is 2.
- The common difference is 3.
- The 2nd term in the sequence is 1.
- To find the 5th term, substitute n = 5 into 3n - 1.

| $n$th term | $3n - 1$ | term |
|---|---|---|
| 1st term ($n = 1$) | $(3 \times 1) - 1$ | 2 |
| 2nd term ($n = 2$) | $(3 \times 2) - 1$ | 5 |
| 3rd term ($n = 3$) | $(3 \times 3) - 1$ | 8 |
| 4th term ($n = 4$) | $(3 \times 4) - 1$ | 11 |

**5.** The $n$th term rule of a sequence is $5n - 3$.
What is the 2nd term in the sequence?

1
2
3

- 7

| $n$th term | $5n - 3$ | term |
|---|---|---|
| 1st term ($n = 1$) | $(5 \times 1) - 3$ | 2 |
| 2nd term ($n = 2$) | $(5 \times ?) - 3$ | |

**6.** The $n$th term rule for a sequence is $20 - 3n$.
Find the 1st term in the sequence.

1
2
3

- 17

| $n$th term | $20 - 3n$ | term |
|---|---|---|
| 1st term ($n = 1$) | $20 - (3 \times ?)$ | |
| 2nd term ($n = 2$) | $20 - (3 \times 2)$ | 14 |

**7.** A sequence has the *n*th term rule $4n + 3$.

$\begin{smallmatrix}1\\2\\3\end{smallmatrix}$ What is the 5th term in this sequence?

▪ 23

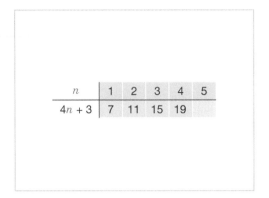

**8.** The *n*th term rule for a sequence is $14 - 2n$.

$\begin{smallmatrix}1\\2\\3\end{smallmatrix}$ Find the 3rd term in the sequence.

▪ 8

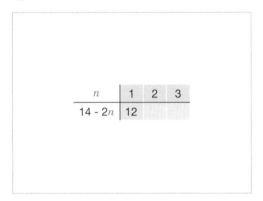

**9.** Find the value of the 7th term in the sequence $6n - 1$.

$\begin{smallmatrix}1\\2\\3\end{smallmatrix}$ ▪ 41

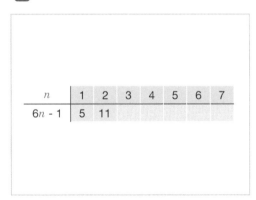

**10.** The *n*th term rule for a sequence is $3n + 2$.

$\begin{smallmatrix}1\\2\\3\end{smallmatrix}$ Find the 9th term in the sequence.

▪ 29

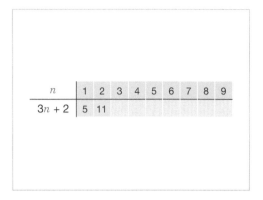

**Level 2:** Fluency - Using the nth term rule for more complex sequences and to find the position of terms.

✱ **Required:** 7/10    ✱ **Student Navigation:** on

✱ **Randomised:** off

**11.** Find the value of the 8th term in the sequence $11n - 5$.

$\begin{smallmatrix}1\\2\\3\end{smallmatrix}$ ▪ 83

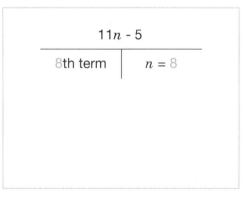

**12.** A sequence has the *n*th term $3 + 7n$.

$\begin{smallmatrix}1\\2\\3\end{smallmatrix}$ What is the value of the 12th term in this sequence?

▪ 87

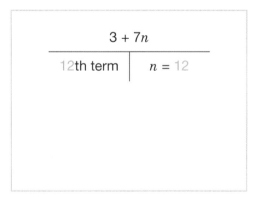

**13.** Find the value of the 6th term in the sequence  with a position to term rule $-2n + 23$.

▪ 11

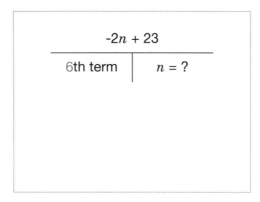

$$\begin{array}{c|c} \multicolumn{2}{c}{-2n + 23} \\ \hline \text{6th term} & n = ? \end{array}$$

**14.** The *n*th term of a sequence is $5n - 7$.  Which term has the value 53?

1/4

▪ 10th ▪ 11th ▪ **12th** ▪ 13th

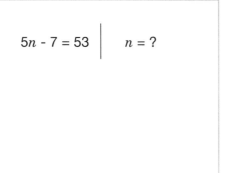

$$\begin{array}{c|c} 5n - 7 = 53 & n = ? \end{array}$$

**15.** A sequence has the *n*th term rule $3n + 5$.  Which term has the value 23?

a
b
c

▪ 6 ▪ **6th**

$$\begin{array}{c|c} 3n + 5 = 23 & n = ? \end{array}$$

**16.** A sequence has the *n*th term rule $61 - 4n$.
a
b
c
Which term has the value 29?

▪ **8th** ▪ 8

**17.** The *n*th term rule of a sequence is $7n - 5$.
a
b
c
Which term has the value 58?

▪ **9th** ▪ 9

**18.** Find the value of the 6th term in the sequence $6n -$  8.

▪ 28

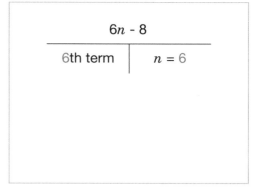

$$\begin{array}{c|c} \multicolumn{2}{c}{6n - 8} \\ \hline \text{6th term} & n = 6 \end{array}$$

**19.** Find the value of the 5th term in the sequence $9n$  $+ 12$.

▪ 57

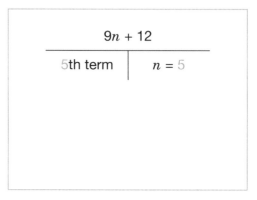

$$\begin{array}{c|c} \multicolumn{2}{c}{9n + 12} \\ \hline \text{5th term} & n = 5 \end{array}$$

**20.** The *n*th term of a sequence is $4n + 7$.  Which term has the value 51?

a
b
c

▪ **11th** ▪ 11

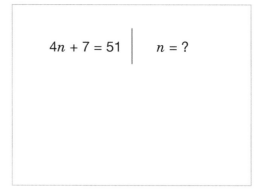

$$\begin{array}{c|c} 4n + 7 = 51 & n = ? \end{array}$$

2/6

**Level 3:** Reasoning - Using the nth term rule.

❋ **Required:** 5/5   ❋ **Student Navigation:** on
❋ **Randomised:** off

**21.** The first five terms of four different sequences are shown in the diagram.
Sort the following *n*th term rules to correctly match the order of sequences.

▪ 19 - n  ▪ 3n + 7  ▪ 20 - 2n  ▪ 12n - 2

| 18 | 17 | 16 | 15 | 14 |
|----|----|----|----|----|
| 10 | 13 | 16 | 19 | 22 |
| 18 | 16 | 14 | 12 | 10 |
| 10 | 22 | 34 | 46 | 58 |

**22.** Jane has calculated the first four terms of the sequence 2*n* + 4 as shown. Explain the mistake Jane has made and calculate the correct answer.

a
b
c

2n + 4

○

2n means the first term will be 2.

+ 4 means that you add 4 to get to the next term.

○ So, the sequence is 2, 6, 10, 14...

**23.** Sort the following *n*th term rules in ascending order (smallest first), according to the value of the **3rd term** in each sequence.

▪ (b)  ▪ (c)  ▪ (a)  ▪ (d)

| (a) 7*n* - 2 | (b) 3*n* + 8 |
|--------------|--------------|
| (c) 30 - 4*n* | (d) 5*n* - 6 |

**24.** Terry says that 54 appears in the sequence 4*n* - 2 .
Is Terry correct? Explain your answer.

a
b
c

**25.** Select two terms that are common to the sequences *3n + 1* and *32 - 5n*.

▪ 7  ▪ 10  ▪ 12  ▪ 13  ▪ 17  ▪ 22

**Level 4:** Problem Solving - Using the nth term rule to find the position of terms.

❋ **Required:** 5/5   ❋ **Student Navigation:** on
❋ **Randomised:** off

**26.** What is the first term that will appear in **both** of the following sequences?
6*n* + 4 and 4*n* - 2.

▪ 10

**27.** Find the 100th term of the sequence with the *n*th term rule 3(*n* + 5).

▪ 315

**28.** The diagram shows a tile pattern for the sequence 4*n* + 2.
What pattern number in this sequence will use exactly 30 tiles?

▪ 7

tiles in the nth term: 4*n* + 2

pattern 1   pattern 2   pattern 3

**29.** A sequence has the *n*th term *50 - 7n*.
What is the first negative term in this sequence?

▪ -6

| *n* | 1 | 2 | ⟶ | *n* |
|-----|---|---|---|-----|
| 50 - 7*n* | | | ⟶ | |

**30.** What is the 7th term of the sequence shown in the diagram.

Give your answer as a decimal.

- 4.25

$n$th term: $\frac{1}{2}n + \frac{3}{4}$

# Number Sequences

**Competency:** Generate and describe linear number sequences.

**Quick Search Ref:** 10299

Correct: Correct.    Wrong: Incorrect. Try again.    Open: Thank you.

**Level 1:** Understanding - Finding numbers in a sequence.

✿ **Required:** 9/12    ✿ **Student Navigation:** on    ✿ **Randomised:** off

**1.** A linear sequence:

1/4

- is the difference between each number in a sequence.
- is used to describe the position of a number in a sequence.
- **is a series of numbers which increase or decrease by the same amount in a sequence.**
- is a letter which represents a number not known in algebra.

**2.** The **difference** in a linear sequence is:

1/4

- **the difference between a term and the next number in a sequence.**
- is used to describe the position of a number in a sequence.
- a set of numbers which increase or decrease by the same amount after each number.
- a letter which represents a number not known in algebra.

**3.** The **position** of a number in a linear sequence is also known as:

1/4

- difference  ▪ equation  ▪ **a term**  ▪ variable

| number | 3 | 6 | 9 | 12 |
| --- | --- | --- | --- | --- |
| position | 1 | 2 | 3 | 4 |

**4.** The formula for this linear sequence is **3n**.

1/4

3, 6, 9, 12...

What is the next number in this sequence?

- ▪ **15**  ▪ 21  ▪ 9  ▪ 30

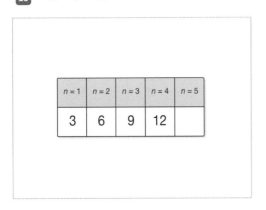

| n = 1 | n = 2 | n = 3 | n = 4 | n = 5 |
| --- | --- | --- | --- | --- |
| 3 | 6 | 9 | 12 | |

**5.** The rule for this linear sequence is **+6**.

3, 9, 15, 21...

Write the number which comes next in this sequence.

- ▪ **27**

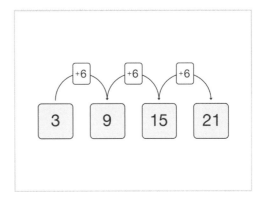

**6.** 25, 20, __, 10, 5.

Write the missing number for this linear sequence.

- ▪ **15**

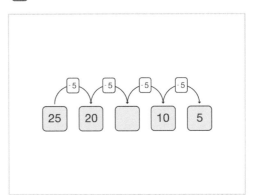

**7.** The rule to find a number in this linear sequence is: **7 × n** (where *n* is the number of the term) **- 1.**

6, 13, 20, 27, 34.

Complete the following:

The **difference** of this sequence is +__ .

▪ **7**

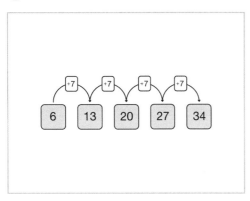

**8.** The rule to find a number in this linear sequence is: **6n** (where *n* is the number of the term).

6, 12, __, 24, 30.

1/4

Select the missing number in the linear sequence.

▪ **18** ▪ **14** ▪ **13** ▪ **23**

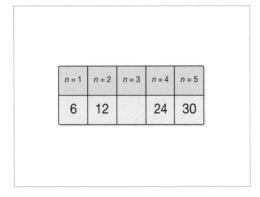

**9.** The rule to find a number for this sequence is: **9n - 5** (where *n* is the number of the term).

4, 13, 22, 31, 40.

The **difference** in this linear sequence is +__.

▪ **9**

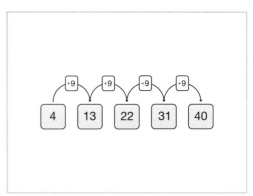

**10.** The formula to find a number in this sequence is: **5n - 2** (where *n* is the number of the term).

3, 8, __, 18, 23.

What is the missing number?

▪ **13**

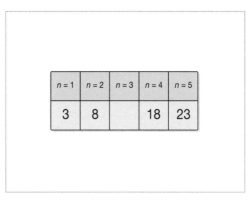

**11.** The **difference** of this sequence is **-7.**

35, 28, __, 14, 7.

Write the **missing number** for this linear sequence.

▪ **21**

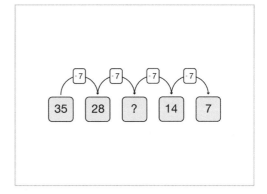

**12.** The rule for this sequence is **+4**.

4, 8, 12, 16...

What number comes next in this linear sequence?

■ 20

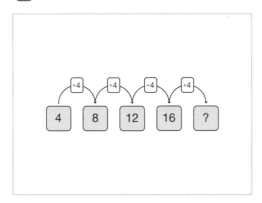

**Level 2:** Fluency - Relating formula and sequence to each other.

✸ **Required:** 7/10    ✸ **Student Navigation:** on
✸ **Randomised:** off

**13.** The difference for this linear sequence is **+6**.

1/5

6, 12, 18, 24...

Which formula matches this sequence?

■ n + 6  ■ 6n  ■ 12n  ■ 18n  ■ 24n

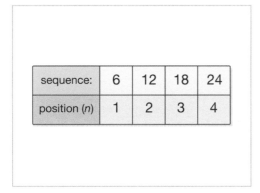

| sequence: | 6 | 12 | 18 | 24 |
|---|---|---|---|---|
| position (n) | 1 | 2 | 3 | 4 |

**14.** The rule to find the next number in this linear sequence is **+2**.

**2** is also **added** once *n* has been multiplied by the difference.

1/5

4, 6, 8, 10...

Choose the formula which matches this sequence.

■ n + 2  ■ n + 3  ■ 2n + 2  ■ 3n  ■ 4n

**15.** The formula to find a number in this sequence is:
**3n - 2**.

1, 4, 7, 10...

What is the **15th term** of this sequence?

■ 43

**16.** The formula for the following sequence is **2n**.

1/4

2, 4, 6, 8...

What is the 20th number in this sequence?

■ 20  ■ 22  ■ 40  ■ 48

**17.** 5, 10, 15, 20...

a
b
c

Write the formula to find **any** term in this sequence by **multiplying *n*** by the **difference**.

■ 5n

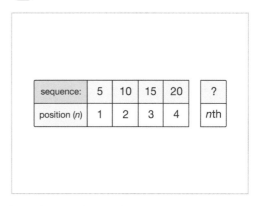

| sequence: | 5 | 10 | 15 | 20 | ? |
|---|---|---|---|---|---|
| position (n) | 1 | 2 | 3 | 4 | nth |

**18.** The difference for this linear sequence is **+4**.

**1** is **subtracted** after *n* has been multiplied by the difference.

1/5

3, 7, 11, 15...

Select the expression which helps you to find the *n*th (any number) term of the sequence.

■ 4n - 1  ■ 3n  ■ n + 2  ■ 2n + 3  ■ n + 4

**19.** 8, 16, 24, 32...

a
b
c

Write the expression which helps you to find the *n*th (any number) term of the sequence by **multiplying *n*** by the **difference**.

■ 8n

**20.** The difference for this linear sequence is +3.

1/5

3, 6, 9, 12...

Which formula can be used to find any number in this sequence?

■ 12n  ■ 9n  ■ 6n  ■ 3n  ■ n + 3

**Level 2:** *cont.*

**21.** The **difference** for this linear sequence is **+2**.

1/4

1 is **added** after *n* has been multiplied by the difference.

3, 5, 7, 9...

Which formula can be used to find any number in this sequence?

■ 2n + 1 ■ 3n ■ n + 2 ■ 3n - 1

**22.** The formula to find any number in the following sequence (the *n*th term) is:

1/4

*n*th term = 5*n*.

5, 10, 15, 20...

What is the **12th** number in this sequence?

■ 12 ■ 32 ■ 60 ■ 80

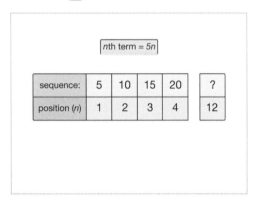

*n*th term = 5*n*

| sequence: | 5 | 10 | 15 | 20 | ? |
|---|---|---|---|---|---|
| position (*n*) | 1 | 2 | 3 | 4 | 12 |

**Level 3:** Reasoning - Writing formulae for sequences with an additional function.

✹ Required: 6/8    ✹ Student Navigation: on
✹ Randomised: off

**23.** Write a formula which helps you to find any number (the *n*th term) in the following linear sequence:

abc

6, 9, 12, 15...

■ 3n+3

**24.** Which formula can be used to calculate the values in the table?

1/4

■ b = a/2 ■ b = a + 2 ■ b = a² ■ b = 2a

| a | b |
|---|---|
| 2 | 4 |
| 4 | 8 |
| 6 | 12 |
| 8 | 16 |

**25.** Is it possible to find the **10th term** in this sequence? Explain your answer.

2, 4, 7, 10, 12, 14...

**26.** Select the first **four** numbers in the sequence for the formula: **8*n* + 3.**

4/7

■ 8 ■ 11 ■ 16 ■ 19 ■ 24 ■ 27 ■ 35

**27.** Write an expression which applies to the following linear sequence:

abc

5, 12, 19, 26...

■ 7n - 2

**28.** Which formula can be used to calculate the values in the table?

1/4

■ b = a - 2 ■ b = a + 2 ■ b = a/2 ■ b = 2a

| a | b |
|---|---|
| 4 | 6 |
| 5 | 7 |
| 6 | 8 |
| 7 | 9 |

**29.** The expressions *n* + **3** and **3*n* + 1** can both be applied to the following sequence:

abc

4, 7, 10, 13, 16, 19.

Is this statement true? Explain your answer.

**30.** Select the first **four** numbers in the sequence for the formula: **6*n* - 5.**

4/7

■ 1 ■ 6 ■ 7 ■ 12 ■ 13 ■ 18 ■ 19

**Level 4:** Problem Solving - Worded questions, negative sequences and tables.

✻ Required: 6/6   ✻ Student Navigation: on
✻ Randomised: off

**31.** Write a formula for the following sequence:

a
b
c

5, -1, -7, -13...

■ -6n+11  ■ 11-6n

**32.** Danny is given £5 by his grandad which he puts into a jar.

a
b
c

His mum puts £2 in the jar every day during January.  On the **1st of January** Danny has £7.

Use a formula to calculate how many pounds are in the jar at the end of January.

*Remember to include the £ sign in your answer.*

■ £67

**33.** A straight line is drawn on a line graph.

a
b
c

Some of the coordinate points of x and y are shown in the table.

Write an equation for the line in terms of x (starting with the x).

■ x=y-1

| x | y |
|---|---|
| 2 | 3 |
| 3 | 4 |
| 4 | 5 |
| 5 | 6 |

**34.** Katie's mum and dad put £200 in a bank account for her on her first birthday.

a
b
c

On her birthday each year, they put an additional £50 in the account.

Calculate how much money Katie will have when she is twenty.

■ £1,150  ■ £1150

**35.** Write a formula that will find any number in the following sequence:

a
b
c

5, 3, 1, -1, -3...

■ 7-2n  ■ -2n+7

**36.** A straight line is drawn on a line graph.

a
b
c

Some of the coordinate points of x and y are shown in the table.

Write an equation for the line in terms of y (starting with y).

■ y=x-3

| x | y |
|---|---|
| 5 | 2 |
| 7 | 4 |
| 9 | 6 |
| 11 | 8 |

# Recognise Sequences - Term to Term

**Competency:**    Generate terms of a sequence from either a term to term or a position to term rule. Recognise geometric sequences and appreciate other sequences that arise.

**Quick Search Ref:**    10080

**Correct:** Correct.    **Wrong:** Incorrect, try again.    **Open:** Thank you.

**Level 1:**    Understanding - Find next and missing terms in arithmetic and geometric sequences.

✳ **Required:** 7/10        ✳ **Student Navigation:** on        ✳ **Randomised:** off

**1.**    Arrange the following terms in the same order as the definitions shown.

■ Arithmetic sequence  ■ Sequence
■ Term-to-term rule  ■ Term  ■ Geometric sequence

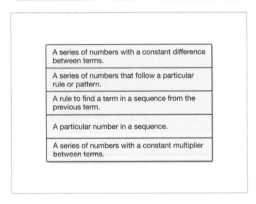

**2.**    What is the next term in the following sequence?
5, 8, 11, 14, ___

 ■ 17

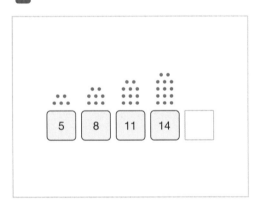

**3.**    Which sequence has a first term of 5 and a term-to-term rule of add 7?

1/4    ■ 5, 10, 15, 20, 25, …  ■ 5, 7, 9, 11, 13, …
■ 5, 12, 19, 26, 33, …  ■ 7, 14, 21, 28, 35, …

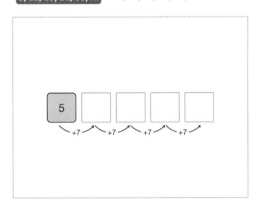

**4.**    What is the next term in this sequence?
3, 6, 12, 24, 48, ___

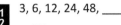 ■ 96

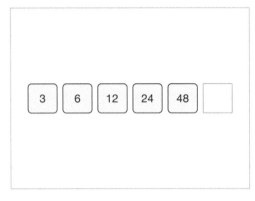

**5.**    What is the missing term in the following sequence?
17, ___, 5, -1, -7

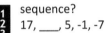 ■ 11

**6.** Which statement describes this sequence?

-2, 1, 4, 7, 10

1/4

- The first term is -2. The term-to-term rule is add 4.
- The fourth term is 10. The term-to-term rule is add 3.
- The first term is -2. The term-to-term rule is subtract 3.
- The third term is 4. The term-to-term rule is add 3.

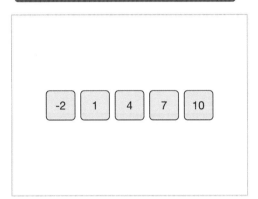

**7.** What is the missing term in the following sequence?

24, 12, 6, 3, ___, 0.75

*Give your answer as a decimal.*

- 1.5

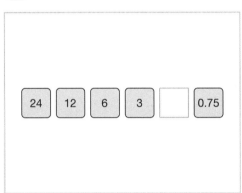

**8.** What is the missing term in this sequence?

-8, ___, 2, 7, 12

- -3

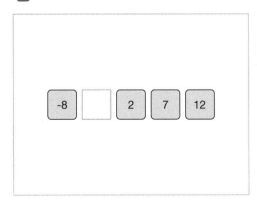

**9.** Which sequence has a fourth term of 11 and a term-to-term rule of subtract 2?

- 5, 7, 9, 11, 13   ■ 11, 9, 7, 5, 3.   ■ 17, 15, 13, 11, 9

1/4   ■ 20, 17, 14, 11, 8

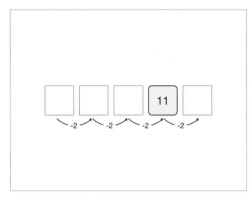

**10.** What is the next term in this sequence?

0.4, 2, 10, 50, ___

- 250

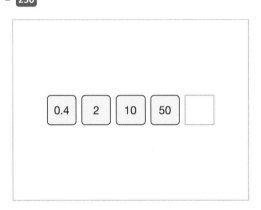

**11.** What is the next term in this sequence of numbers?

1, 4, 9, 16, ___

- 25

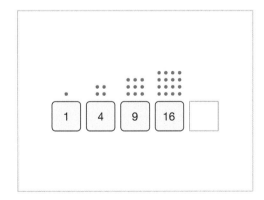

**Level 2:** *cont.*

**12.** The following sequence is called the Fibonacci sequence. What is the next term?

1, 1, 2, 3, 5, 8, 13, ___

▪ 21

**13.** The first four terms in a sequence are: 5, 8, 11, 14. What is the 7th term?

▪ 23

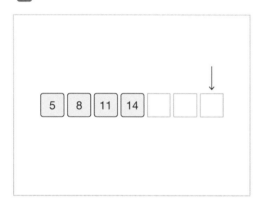

**14.** What is the fifth term in the following sequence if the term-to-term rule is add 1 to the previous term and multiply your answer by 2? 1, 4, 10.

▪ 46

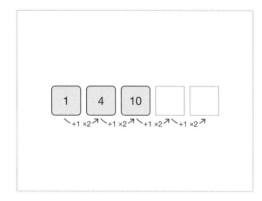

**15.** The series of triangular numbers starts with the numbers 1, 3, 6, 10. What is the next term in the sequence?

▪ 15

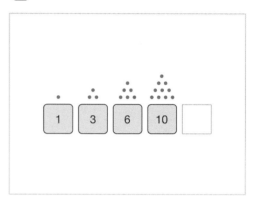

**16.** What is the first term in the following sequence?

___, ___, ___, 24, 48, 96

▪ 3

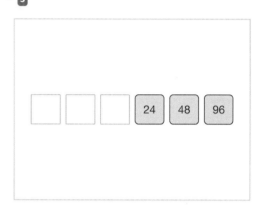

**17.** What is the next term in this sequence?

2, 5, 10, 17, 26, ___

▪ 37

**18.** What is the next term in the following sequence?

2, 6, 12, 20, 30, 42, ___

▪ 56

**19.** What is the next term in the Fibonacci sequence?

3, 4, 7, 11, 18, 29, ___

▪ 47

**Level 2:** *cont.*

**20.** What is the 7th term in the following sequence:
320, 160, 80, 40, ...

**1 2 3** ▪ **5**

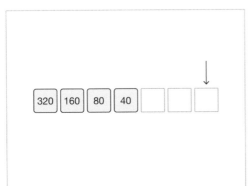

**Level 3:** Reasoning - Misconceptions with sequences and comparing different terms in the same sequence.

✱ **Required:** 5/5    ✱ **Student Navigation:** on
✱ **Randomised:** off

**21.** Gavin is thinking of a sequence with a term-to-term rule of add 3.
Eve says, "The first five terms of your sequence must be 3, 6, 9, 12 and 15".
Is Eve correct? Explain your answer.

**a b c**

**22.** An arithmetic sequence has a first term of -9 and a term-to-term rule of +7.
What is the value of the 65th term minus the 62nd term?

**1 2 3**

▪ **21**

**23.** A sequence has a first term of 5 and a term-to-term rule of multiply by -1.
What is the value of the 100th term in the sequence?

**1 2 3**

▪ **-5**

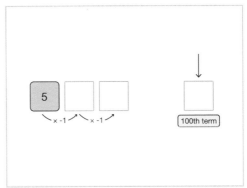

**24.** Katie and Imran are each thinking of arithmetic sequences.
George says 'If I add the corresponding terms of your two sequences, I will get another arithmetic sequence'.
Is George correct? Explain your answer.

**a b c**

**25.** A geometric sequence has a first term of 7 and a term-to-term rule of multiply by 9.
What is the value of the 10th term divided by the 8th term?

**a b c**

▪ **81**

**Level 4:** Problem Solving - Comparing multiple sequences and calculating multiple terms in sequences.

✱ **Required:** 5/5    ✱ **Student Navigation:** on
✱ **Randomised:** off

**26.** A sequence has a term-to-term rule of subtract 7 then multiply by 3. If the third term of the sequence is 96, what is the first term of the sequence?

**1 2 3**

▪ **20**

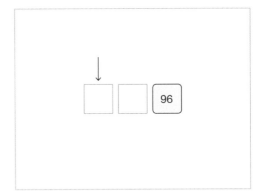

**27.** Sort the following sequences in ascending order according to the value of their fifth terms.

- The first term is 19, the rule is +7.
- The first term is 3, the rule is ×2.
- The first term is 61, the rule is -3.
- The first term is 800, the rule is ÷2.

| 800 | | | | |
| --- | --- | --- | --- | --- |

÷2

| 3 | | | | |
| --- | --- | --- | --- | --- |

×2

| 19 | | | | |
| --- | --- | --- | --- | --- |

+7

| 61 | | | | |
| --- | --- | --- | --- | --- |

-3

**28.** A biologist is studying the population of bees in his area and notices it follows a Fibonacci sequence. If the bee population increased from 5,500 in 2016 to 8,900 in 2017, what was the population in 2008?

- 100

| year | population of bees |
| --- | --- |
| 2008 | |
| 2009 | |
| 2010 | |
| 2011 | |
| 2012 | |
| 2013 | |
| 2014 | |
| 2015 | |
| 2016 | 5,500 |
| 2017 | 8,900 |

**29.** A sequence has a term-to-term rule of subtract 4. If the sum of the first five terms is 45, what is the first term?

- 17

**30.** Sort the sequences in order so they match the rules shown.

- 3, 6, 11, 18, 27, ...
- 3, 6, 12, 24, 48, ...
- 3, 4, 7, 11, 18, 29, ...
- 1, 3, 6, 10, 15, ...

| Square numbers, add 2. |
| --- |
| First term is 3, the rule is ×2. |
| Fibonacci sequence, 1st term 3, 2nd term 4. |
| Triangular numbers. |

# Collect Like Terms (Single Variable)

**Competency:** Simplify and manipulate algebraic expressions to maintain equivalence by collecting like terms.

**Quick Search Ref:** 10249

Correct: Correct.    Wrong: Incorrect, try again.    Open: Thank you.

**Level 1:** Understanding - Collect terms in expressions with a single variable.

❋ **Required:** 7/10         ❋ **Student Navigation:** on         ❋ **Randomised:** off

**1.** Simplify the following expression:

$y + y$.

▪ 2 x y ▪ y2 ▪ $y^2$ ▪ **2y**

1/4

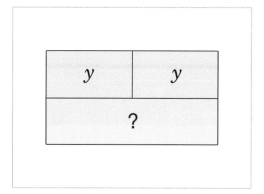

**2.** What is the following expression written as a single term?

$b + b + b + b$.

1/4   ▪ 4 ▪ 4 x b ▪ $b^4$ ▪ **4b**

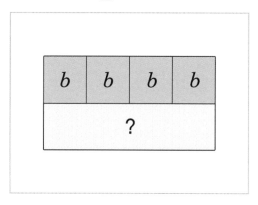

**3.** Simplify:
3t + 2t.

▪ **5t**

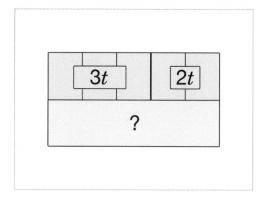

**4.** Give the following as a single term:

$5c + 2c + c$.

▪ **8c**

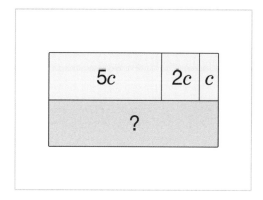

**5.** Simplify the following expression:
7y - 3y.

▪ **4y**

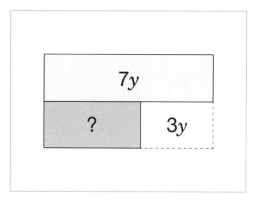

**6.** Simplify the following expression by writing it as a single term:
7g - 3g + 2g.

▪ **6g**

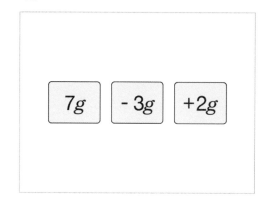

     Ref:10249   Collect Like Terms (Single Variable)

**Level 1:** *cont.*

**7.** Simplify:

$3v - 8v$.

■ -5v

**8.** Simplify the following:

$4d + d + 3d$.

■ 8d

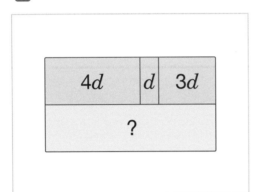

**9.** Give the following as a single term:

$8n - 3n + 4n$.

■ 9n

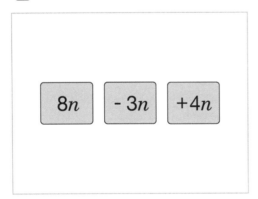

**10.** Simplify:

$2x - 6x$.

■ -4x

**Level 2:**   Fluency - Collect like terms in expressions with a variable and constant terms.

✱ **Required:** 7/10    ✱ **Student Navigation:** on
✱ **Randomised:** off

**11.** Which is the constant term in the following expression?

$5b + 7 - 2b$.

1/4   ■ 5  ■ b  ■ 7  ■ -2

$$5b + 7 - 2b$$

**12.** Which **two terms** contain variables in the following?

$7m + 9 - 2m - 5$.

1/4   ■ 7m and -2m  ■ 7 and -2  ■ 9 and -5  ■ -2m and -5

$$7m + 9 - 2m - 5$$

**Level 2:** *cont.*

**13.** Simplify the following expression:
 4z + 5z + 2.

■ 11z ■ 9z + 2 ■ 11 ■ 2z + 2

1/4

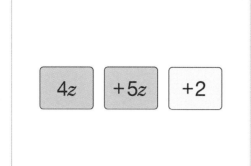

**14.** Simplify by collecting like terms:
**a**
**b** 9a + 3 + 6a + 4.
**c** ■ 15a + 7 ■ 7 + 15a

**15.** Simplify:
**a**
**b** 5k + 7 + 2k - 4.
**c** ■ 7k + 3

**16.** Collect like terms in the following expression:
**a** 6p + 11 - 2p + 4.
**b**
**c** ■ 15 + 4p ■ 4p + 15

**17.** Simplify the following expression:
**a**
**b** 7y + 2 - 5y - 11.
**c** ■ 2y - 9

$$7y + 2 - 5y - 11$$

**18.** Simplify:
**a**
**b** 3d - 2 + 5d + 8.
**c** ■ 8d + 6 ■ 6 + 8d

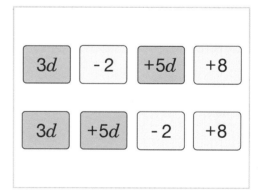

**Level 2:** *cont.*

**19.** Collect like terms in the following expression:
**a b c** $4t + 1 - 2t - 6$.

■ 2t - 5

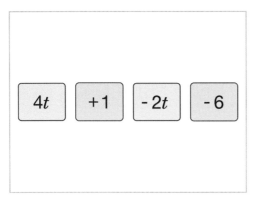

**20.** Simplify the following by collecting like terms:
**a b c** $2e + 3 - 5e + 4$.

■ - 3e + 7  ■ 7 - 3e

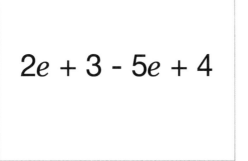

**Level 3:** Reasoning - Adding and subtracting expressions with constant terms and terms involving one variable.

✹ **Required:** 5/5   ✹ **Student Navigation:** on
✹ **Randomised:** off

**21.** What is the missing term?
**a b c** $13h + ? = 18h$.

■ 5h

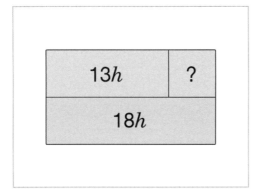

**22.** Luke says that $4r + 5 + 2r = 11r$. Is Luke correct?
**a b c** Explain your answer.

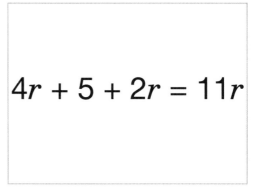

**23.** Write a simplified expression for the perimeter of
**a b c** the triangle.

■ 9v - 3

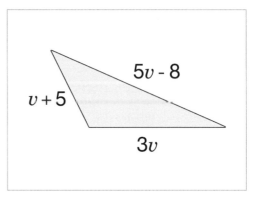

**24.** What missing expression makes the equation
**a b c** balance?
$5x + 8 + ? = 7x + 2$

■ 2x - 6

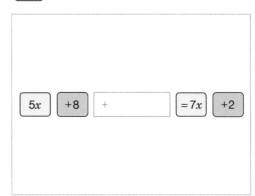

### Level 3: *cont.*

**25.** I think of three consecutive numbers. The smallest
**a** equals *n*. Write an expression for the sum of the
**b** three numbers.
**c**

▪ 3 + 3n ▪ 3n + 3

---

### Level 4: Problem Solving - Problem solving involving expressions with constant terms and terms involving one variable.

✳ **Required:** 5/5   ✳ **Student Navigation:** on
✳ **Randomised:** off

**26.** The sum of two side-by-side bricks is the answer
**a** to the brick above.
**b** What expression goes in the top brick of the
**c** pyramid?

▪ 15x + 2

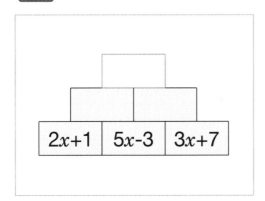

$2x+1$   $5x-3$   $3x+7$

**27.** The grid shows nine expressions. There are four
**a** pairs of equivalent expressions. Which is the odd
**b** one out? Simplify your answer.
**c**

▪ 2 + 2a ▪ 2a + 2

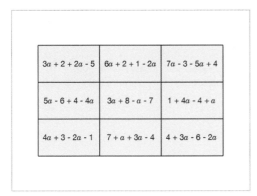

| $3a + 2 + 2a - 5$ | $6a + 2 + 1 - 2a$ | $7a - 3 - 5a + 4$ |
| $5a - 6 + 4 - 4a$ | $3a + 8 - a - 7$ | $1 + 4a - 4 + a$ |
| $4a + 3 - 2a - 1$ | $7 + a + 3a - 4$ | $4 + 3a - 6 - 2a$ |

**28.** The expression in a circle is equal to the sum of
**a** the expressions in the squares on either side of it.
**b** What expression goes in the empty circle?
**c**

▪ 9x - 3

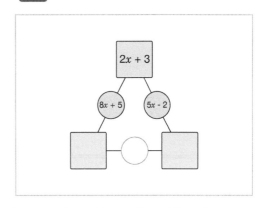

$2x + 3$

$8x + 5$   $5x - 2$

**29.** The sum of two side-by-side bricks is the answer
**a** to the brick above.
**b** What expression should go in the striped brick of
**c** this algebraic pyramid?

▪ 3z + 5

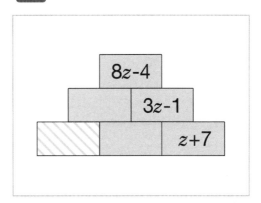

$8z-4$

$3z-1$

$z+7$

**30.** The perimeter of the triangle is equal to the
**1** perimeter of the square, so what is the value of
**2** *w*?
**3**

▪ 9

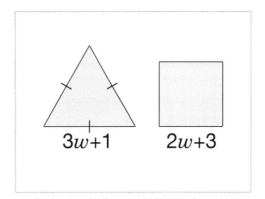

$3w+1$      $2w+3$

---

# Collect Like Terms (Two or More Variables)

**Competency:** Simplify and manipulate algebraic expressions to maintain equivalence by: collecting like terms.

**Quick Search Ref:** 10305

Correct: Correct.    Wrong: Incorrect, try again.    Open: Thank you.

**Level 1:** Understanding - Collecting like terms in expressions with two or more different variables.

✿ **Required:** 7/10    ✿ **Student Navigation:** on    ✿ **Randomised:** off

---

**1.** Simplify the following expression by collecting like terms:

$a + a + a + b + b$

1/4    ■ 5ab  ■ a³ + b²  ■ 3a + 2b  ■ 3 x a + 2 x b

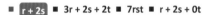
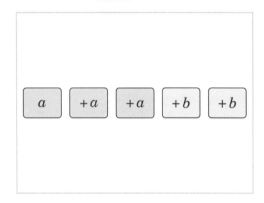

**2.** Simplify:
$r + s + r + t - r + s - t$

■ r + 2s  ■ 3r + 2s + 2t  ■ 7rst  ■ r + 2s + 0t

1/4

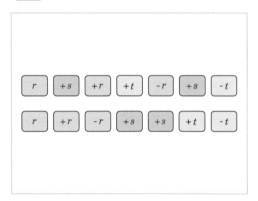

**3.** Collect like terms in the following expression:
a
b
c
$i - j - k + i + k + k - i + i - j$

■ k + 2i - 2j  ■ -2j + 2i + k  ■ k - 2j + 2i  ■ 2i - 2j + k
■ 2i + k - 2j  ■ -2j + k + 2i

**4.** Simplify the following:
a
b
c
$3u + 5v + 6u + 2v$

■ 9u + 7v  ■ 7v + 9u

**5.** Simplify:
a
b
c
$6e + 7f + 5e - 2f$

■ 11e + 5f  ■ 5f + 11e

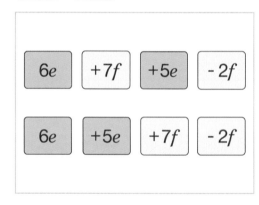

**6.** Collect like terms:
a
b
c
$4m - 3n + 3m - n$

■ 7m - 4n

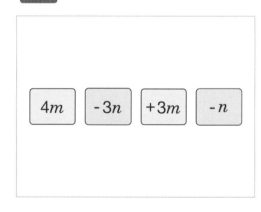

**Level 1:** *cont.*

**7.** Simplify by collecting like terms:

**a**
**b**   $3a + 2b - 5c - b + 2a + 4c$
**c**

■ $b + 5a - c$   ■ $5a - c + b$   ■ $b - c + 5a$   ■ $5a + b - c$
■ $- c + 5a + b$   ■ $- c + b + 5a$

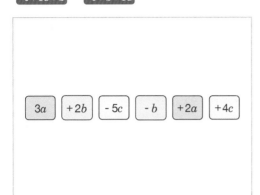

**8.** Simplify the following:

**a**
**b**   $7x - 5y - 2x + 3y$
**c**

■ $5x - 2y$

**9.** Simplify:

**a**
**b**   $6g + 7h - 2g + 4h$
**c**

■ $4g + 11h$   ■ $11h + 4g$

**10.** Simplify the following by collecting like terms:

**a**
**b**   $3r - 12s - 5 + r + 9s + 2$
**c**

■ $-3s + 4r - 3$   ■ $- 3 - 3s + 4r$   ■ $4r - 3s - 3$   ■ $4r - 3 - 3s$
■ $-3s - 3 + 4r$   ■ $- 3 + 4r - 3s$

**Level 2:**   Fluency - Collecting like terms in expressions involving indices and products of variables.

✹ **Required:** 7/10   ✹ **Student Navigation:** on
✹ **Randomised:** off

**11.** What is the coefficient of the $y^2$ term in the following expression?

☐
☒   $4y^2 + 5y + 7$
☐

1/4   ■ $4$ ■ 2 ■ 5 ■ 7

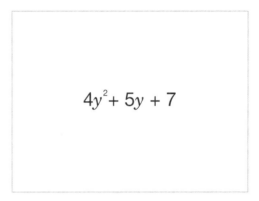

$$4y^2 + 5y + 7$$

**12.** Simplify the following:

☐
☒   $5z^2 + 3 + 4z^2$
☐

1/4   ■ $9z^2 + 3$   ■ $12z^2$ ■ $9z^2$ ■ 12

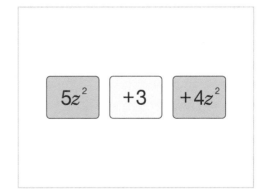

Level 2: *cont.*

**13.** If you collected like terms in the following expression, what would the **coefficient** of the $b^2$ term be?

1/4

$5b^2 + 3b - 2b^2 + 7b$

- 3   - 7   - 13   - $3b^2$

$$5b^2 + 3b - 2b^2 + 7b$$

**14.** What is the coefficient of the *bc* term in the following expression?

$5b + 2bc + 7c$

1/4   - 12   - 2bc   - 14   - 2

$$5b + 2bc + 7c$$

**15.** Simplify the following expression by collecting like terms:

a
b
c

$5m + 7mn + 2m - 4mn$

- 3mn + 7m   - 7m + 3mn

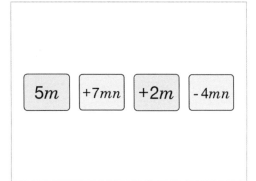

**16.** Simplify the following:

a
b
c

$5rs - 2s + 3sr - s$

- 8sr - 3s   - 8rs - 3s

$$5rs - 2s + 3sr - s$$

**17.** Collect like terms in the following expression:

a
b
c

$7e + 4f - 2ef - 6f - 2e + 5ef$

- 3ef - 2f + 5e   - - 2f + 5e + 3ef   - 3ef + 5e - 2f
- 5e - 2f + 3ef   - 5e + 3ef - 2f   - - 2f + 3ef + 5e

$$7e + 4f - 2ef - 6f - 2e + 5ef$$

**18.** Which of the following expressions simplifies to $5n^2 - 2n$?

1/4

A) $3n^2 - n + 2n^2 + n$
B) $4n^2 - n - n^2 - n$
C) $7n^2 + 4n - 2n - 2n^2$
D) $8n^2 - 5n - 3n^2 + 3n$

- A   - B   - C   - D

**19.** Simplify the following expression:

a
b
c

$7g - 3gh - 2g + 6hg$

- 3hg + 5g   - 5g + 3hg   - 5g + 3gh   - 3gh + 5g

$$7g - 3gh - 2g + 6hg$$

Level 2: *cont*.

**20.** Collect like terms in the following expression:

3*j* + *k* - 2*jk* + 4*j* - 7*k* + 5*jk*

- 7j - 6k + 3jk   ·   -6k + 3jk + 7j   ·   3jk + 7j - 6k
- 7j + 3jk - 6k   ·   3jk - 6k + 7j   ·   -6k + 7j + 3jk

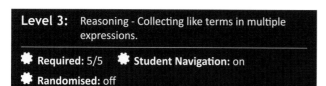

$$3j + k - 2jk + 4j - 7k + 5jk$$

**Level 3:**   Reasoning - Collecting like terms in multiple expressions.

✸ **Required:** 5/5     ✸ **Student Navigation:** on
✸ **Randomised:** off

**21.** Sadie is completing a question on collecting like terms.

Is Sadie correct? Explain your answer.

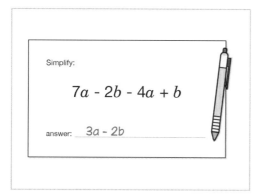

Simplify:

$$7a - 2b - 4a + b$$

answer:   *3a - 2b*

**22.** Write an expression for the perimeter of a triangle with sides of length:

7*a*, 6*a* + *b* and 4*b* - 3*a*.

- 5b + 10a   ·   10a + 5b

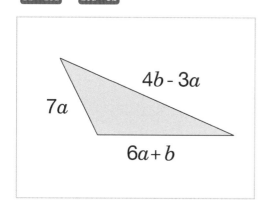

*4b - 3a*

*7a*

*6a + b*

**23.** What is the missing expression which will make the equation correct?

- 2x - 3y - 6

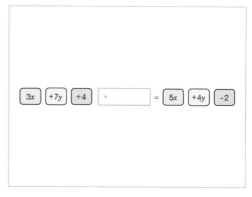

$3x$  $+7y$  $+4$  $+$  [     ]  $=$  $5x$  $+4y$  $-2$

**24.** Ismail is completing a question on collecting like terms.

Is Ismail correct? Explain your answer.

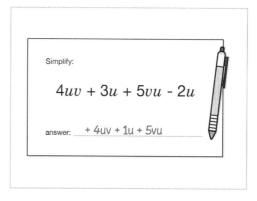

Simplify:

$$4uv + 3u + 5vu - 2u$$

answer:   + 4uv + 1u + 5vu

**25.** Which expression simplifies to $2x^2 + 3$?

- $x^2 + 2 - x^2 + 1$   ·   $x^2 - 1 + x^2 + 2$   ·   $3x^2 + 7 - 5x^2 - 4$
- $4x^2 - 2 - 2x^2 + 5$

1/4

**Level 4:**   Problem Solving - Inverse calculations and multi-step problems.

✸ **Required:** 5/5     ✸ **Student Navigation:** on
✸ **Randomised:** off

**26.** The sum of the two side-by-side bricks is the answer to the brick above. What expression goes in the brick at the top of the pyramid?

- 8r - 2s

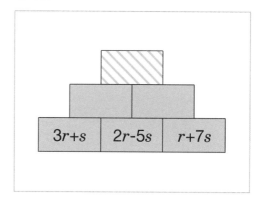

$3r+s$   $2r-5s$   $r+7s$

**27.** The expression in a circle is equal to the sum of the expression in the squares on either side of it. What expression goes in the blank circle?

a
b
c

■ `- v + 3u`  ■ `3u - v`

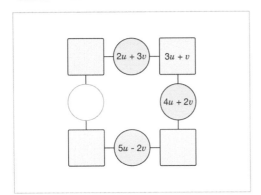

**30.** A cross is made from 5 identical rectangles. If the height of the rectangle is $r$ and the total perimeter of the cross is $18s - 6r$, what is the length of each rectangle?

a
b
c

■ `3s - 2r`

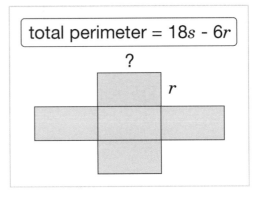

**28.** The grid shows nine expressions. There are four pairs of equivalent expressions. Which is the odd one out? Simplify your answer.

a
b
c

■ `a - b`  ■ `- b + a`

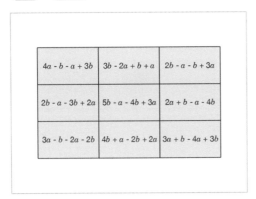

**29.** The sum of the two side-by-side bricks is the answer to the brick above. What expression goes in the brick on the striped box on the bottom row?

a
b
c

■ `4b - 3a`

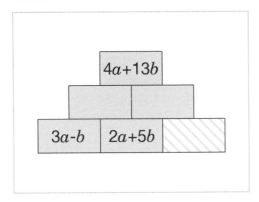

# Divide Terms

**Competency:** Simplify and manipulate algebraic expressions to maintain equivalence.

**Quick Search Ref:** 10221

Correct: Correct.    Wrong: Incorrect, try again.    Open: Thank you.

**Level 1:** Understanding - Dividing terms with a single variable.

❋ Required: 7/10            ❋ Student Navigation: on            ❋ Randomised: off

1.  What is $6a \div 2$?

    a b c  ▪ 3a

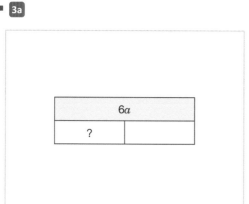

2.  Simplify $18m \div 3$.

    a b c  ▪ 6m

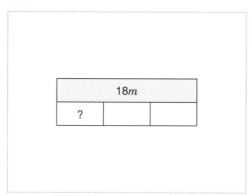

3.  What is the highest common factor of $16d$ and $12d$?

    a b c  ▪ 4d

4.  What is $5k \div k$?

    a b c  ▪ 5

$$\frac{5k}{k}$$

5.  Simplify $28w \div 14w$.

    a b c  ▪ 2

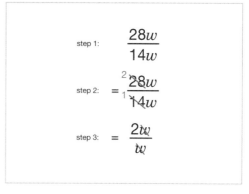

step 1: $\dfrac{28w}{14w}$

step 2: $= \dfrac{^{2}\cancel{28}w}{_{1}\cancel{14}w}$

step 3: $= \dfrac{2\cancel{w}}{\cancel{w}}$

6.  What is $9t^2 \div t$?

    a b c  ▪ 9t

$$\frac{9t^2}{t}$$

$$= \frac{9 \times t \times t}{t}$$

7.  Simplify $42q^2 \div 7q$.

    a b c  ▪ 6q

$$\frac{42q^2}{7q}$$

**Level 1:** *cont.*

**8.** Simplify $80g \div 16g$.

 ▪ 5

$$\frac{80g}{16g}$$

**9.** What is $11r^2 \div r$?

 ▪ 11r

$$\frac{11r^2}{r}$$

$$= \frac{11 \times r \times r}{r}$$

**10.** Simplify $72h^2 \div 9h$.

 ▪ 8h

$$\frac{72h^2}{9h}$$

**11.** What is $7ab \div a$?

 ▪ 7b

$$\frac{7ab}{a}$$

**12.** Simplify $24mn \div 3n$.

 ▪ 8m

$$\frac{24mn}{3n}$$

**13.** What is $48x^2y \div 6x$?

 ▪ 8xy

$$\frac{48x^2y}{6x}$$

$$= \frac{48 \times x \times x \times y}{6 \times x}$$

**Level 2:** *cont.*

**14.** Simplify $3c^2 \div 6c$.

 ▪ c/2

$$\frac{3c^2}{6c}$$

$$= \frac{3 \times c \times c}{6 \times c}$$

**15.** What is $30uvw \div 6uv$?

 ▪ 5w

$$\frac{30uvw}{6uv}$$

**16.** Simplify $52a^2bc \div 13ab$.

 ▪ 4ac

$$\frac{52a^2bc}{13ab}$$

**17.** Which is the correct simplification of $18rst^2 \div 27st$?

 ▪ (i) ▪ (ii) ▪ (iii) ▪ (iv)

1/4

$$\frac{18rst^2}{27st}$$

(i) $\frac{rt}{3}$  (ii) $\frac{2r}{3}$  (iii) $\frac{2rt}{3}$  (iv) $\frac{2rt^2}{3}$

**18.** Simplify $56ijk \div 8ik$.

 ▪ 7j

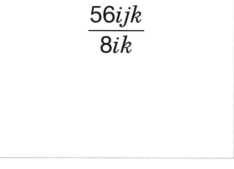

$$\frac{56ijk}{8ik}$$

**19.** What is $63m^2n \div 7m$?

 ▪ 9mn

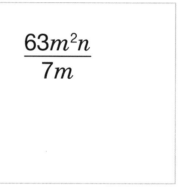

$$\frac{63m^2n}{7m}$$

**20.** Simplify $5x^2y \div 15x$.

 ▪ xy/3 ▪ yx/3

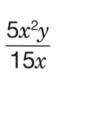

$$\frac{5x^2y}{15x}$$

**Level 3:** Reasoning - Misconceptions and applying
dividing terms to problems.

✸ **Required:** 5/5    ✸ **Student Navigation:** on
✸ **Randomised:** off

**21.** The following are the first four terms of a
**a** geometric sequence:
**b**
**c** $1, 2a, 4a^2, 8a^3$
What is the value of the 10th term in the
sequence divided by the 9th term?

▪ **2a**

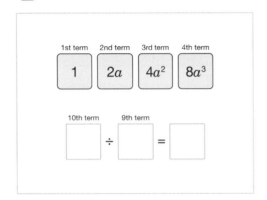

**22.** Simplify the fraction shown.
**a**
**b**  ▪ **3a + 4b**
**c**

$$\frac{6a^2 + 8ab}{2a}$$

**23.** Jasper made one mistake in his homework on
**a** dividing terms. What is the correct answer to the
**b** question he got wrong?
**c**

▪ **m/2**

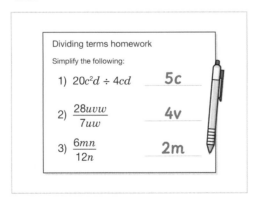

**24.** What is the missing denominator?
**a**
**b** ▪ **8a**
**c**

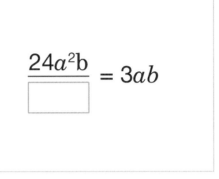

$$\frac{24a^2b}{\boxed{\phantom{xxx}}} = 3ab$$

**25.** Suleman says that $x \div x$ is 0.
**a** Is he correct? Explain your answer.
**b**
**c**

**Level 4:** Problem Solving - Multi-step problems
involving dividing terms.

✸ **Required:** 5/5    ✸ **Student Navigation:** on
✸ **Randomised:** off

**26.** Complete the multiplication grid. What expression
**a** goes in the striped box?
**b**
**c** ▪ **42ab**

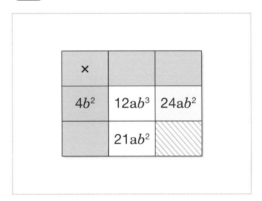

**27.** If the shape shown has an area of *78jk,* what is the
shape's height?

a
b
c
 ▪ 9j

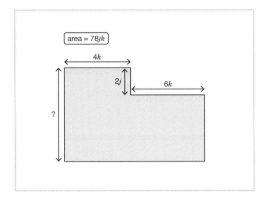

**28.** The grid contains four pairs of equivalent
expressions. Which expression is the odd one out?

a
b
c
 ▪ 3n

| $\dfrac{36mn}{4m}$ | $6m$ | $24n \div 3$ |
|---|---|---|
| $8n$ | $12mn^2 \div 3n^2$ | $3n$ |
| $\dfrac{42m^2n}{7mn}$ | $9n$ | $4m$ |

**29.** The term in each circle is equal to the product of
the terms in the squares on either side of it. What
term goes in the blank circle?

a
b
c

 ▪ 63sr  ▪ 63rs

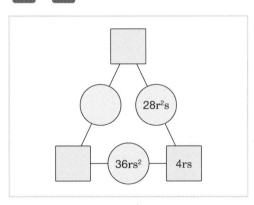

**30.** The square and the triangle have the same area.
What is the length of the base of the triangle?

a
b
c
 ▪ 18ab

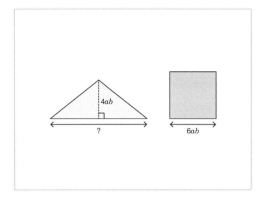

# Multiply Terms

**Competency:** Simplify and manipulate algebraic expressions to maintain equivalence by: multiplying terms.

**Quick Search Ref:** 10096

Correct: Correct.    Wrong: Incorrect, try again.    Open: Thank you.

**Level 1:** Understanding - Multiplying simple terms.

✹ **Required:** 7/10     ✹ **Student Navigation:** on     ✹ **Randomised:** off

---

**1.** Give the following as a single term:

$4 \times a$

▪ **4a**

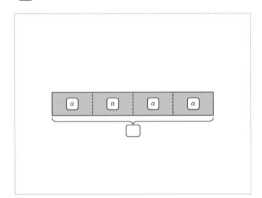

**2.** Simplify:

$r \times 7$

▪ **7r**

**3.** Multiply the following terms:

$3 \times 2w$

▪ **6w**

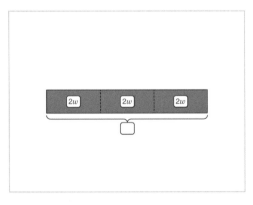

**4.** Simplify the following:

$7 \times 8g$

▪ **56g**

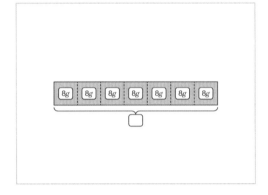

**5.** Give the following as a single term:

$3t \times 5$

▪ **15t**

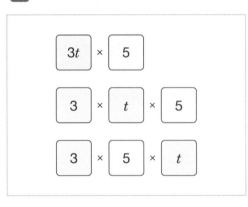

**6.** Simplify:

$6c \times 4d$

▪ **24dc**  ▪ **24cd**

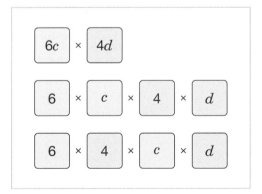

---

182                    Ref:10096    Multiply Terms

**Level 1:** *cont.*

7.  Multiply the following two terms:
    a
    b
    c
    $8n \times 5m$
    ▪ 40mn  ▪ 40nm

8.  Simplify the following:
    a
    b
    c
    $9f \times 6$
    ▪ 54f

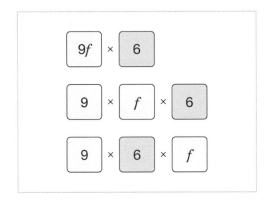

9.  Multiply the following terms:
    a
    b
    c
    $7r \times 6s$
    ▪ 42rs  ▪ 42sr

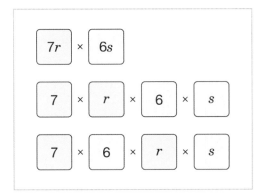

10. Simplify the following:
    a
    b
    c
    $8v \times 4u$
    ▪ 32vu  ▪ 32uv

**Level 2:**  Fluency - Multiplying 3 terms and those involving indices.

✱ **Required:** 7/10    ✱ **Student Navigation:** on
✱ **Randomised:** off

11. Simplify the following:
    a
    b
    c
    $4r \times 2s \times t$
    ▪ 8srt  ▪ 8str  ▪ 8trs  ▪ 8tsr  ▪ 8rts  ▪ 8rst

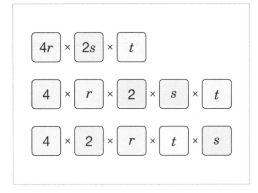

12. Give the following expression as a single term:
    a
    b
    c
    $2d \times e \times 9f$.
    ▪ 18fed  ▪ 18dfe  ▪ 18def  ▪ 18efd  ▪ 18fde
    ▪ 18edf

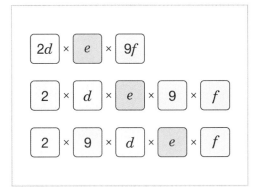

13. Multiply the following terms:
    a
    b
    c
    $3a \times 5b \times 4c$
    ▪ 60acb  ▪ 60bac  ▪ 60bca  ▪ 60cab  ▪ 60abc
    ▪ 60cba

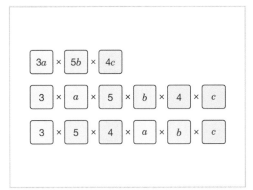

Level 2: *cont.*

**14.** What is the following expression in its simplest form?

$7t \times 4s \times 5r$

- 140rts
- 140trs
- 140rst
- 140srt
- 140str
- 140tsr

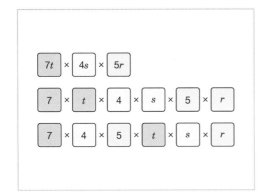

**15.** Simplify the following:

$3g \times 8g$

- 24gg
- 24g
- 48g
- $24g^2$

1/4

**16.** What is the product of the following terms?

$7u \times 9uv$

- 63uv
- $63u^2v$
- $63u^2v^2$
- $63uv^2$

1/4

**17.** Simplify:

$2a^2b \times 3ac \times 4b$

- $24a^3b^2c$
- $24a^2b^2c$
- 24abc
- 24aaabbc

1/4

**18.** Give the following as a single term:

$g \times 7h \times 3f$

- 21ghf
- 21hgf
- 21gfh
- 21fhg
- 21hfg
- 21fgh

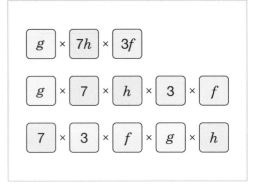

**19.** Multiply the following terms:

$7s \times 3r \times 6t$

- 126rst
- 126rts
- 126tsr
- 126trs
- 126srt
- 126str

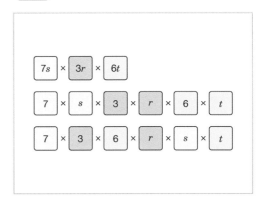

**20.** Multiply the following terms:

$9a^2 \times 5ab$

- 45ab
- $45a^2b$
- $45a^2b^2$
- $45a^3b$

1/4

**21.** Marcello is simplifying an expression by collecting like terms.

Is Marcello correct? Explain your answer.

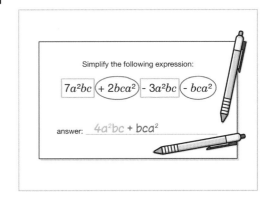

Simplify the following expression:

$7a^2bc + 2bca^2 - 3a^2bc - bca^2$

answer:   $4a^2bc + bca^2$

**22.** If $8r \times ? = 72r^2s$, find the value of the missing term.

- 9sr
- 9rs

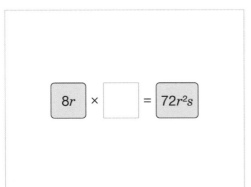

**23.** Write an expression for the area of a triangle with base length $12e$ and perpendicular height $3f$.

- 18fe
- 18ef

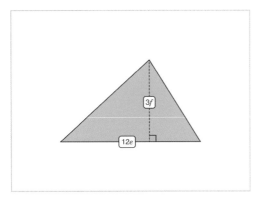

**24.** Madison has completed her maths homework. Which answer did Madison get wrong?

1/4
- $7q \times 5p \times 3r = 105qpr$
- $4a^2 \times 9ab = 36a^3b$
- $6u \times 7vw \times 4uw = 168u^2vw$
- $9sr \times 12r = 108sr^2$

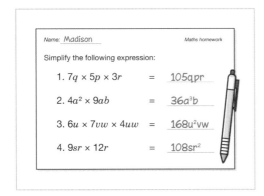

**25.** A square has an area of $64d^2$.
What is the length of one side of the square?

a
b
c
- 8d

---

**Level 4:**  Problem Solving - Multiplying several variables and inverse questions.

� **Required:** 5/5   ✱ **Student Navigation:** on
✱ **Randomised:** off

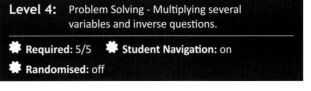

**26.** In the algebraic pyramid, the term in each brick is the product of the terms in the two bricks directly below it. What term goes in the striped brick?

a
b
c
- 5t

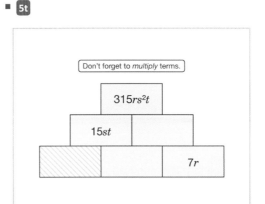

**27.** The diagram shows a large rectangle with a smaller rectangle removed along its top edge. Give an expression for the area of the remaining shape.

a
b
c
- 6cd
- 6dc

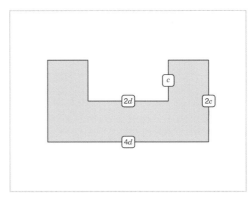

**28.** The term of in each square is the product of the two circles on either side it.
What term goes in the blank square?

a
b
c
- 21vu
- 21uv

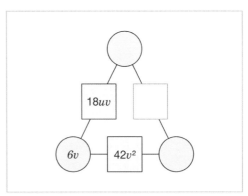

---

**Level 4: cont.**

**29.** A large square is made from a smaller square and
**a**  4 identical rectangles.
**b**  The perimeter of the large square is 48*p* and its
**c**  area is 4 times that of the smaller square. What is
     the length of a rectangle?

■ 9p

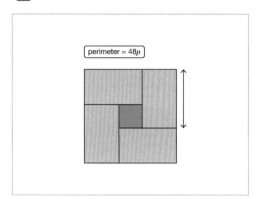

**30.** Complete the multiplication grid.
**a**  What term goes in the striped box?
**b**
**c**  ■ 12ba  ■ 12ab

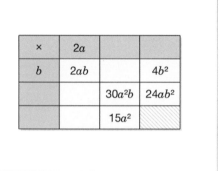

# Use Algebraic Notation

**Competency:** Use and interpret algebraic notation including: ab in place of a × b, 3y in place of y + y + y and 3 × y, a² in place of a × a, a³ in place of a × a × a, a²b in place of a × a × b, a/b in place of a ÷ b and coefficients written as fractions.

**Quick Search Ref:** 10301

**Correct:** Correct.　　**Wrong:** Incorrect, try again.　　**Open:** Thank you.

---

**Level 1:** Understanding - Definitions and recognising notation.

✿ **Required:** 10/10　　✿ **Student Navigation:** on　　✿ **Randomised:** off

---

1. Arrange the definitions in the following order:

   term
   variable
   coefficient

   - A single number or variable, or numbers and variables multiplied together.
   - A varying quantity that is represented by a letter or symbol.
   - A number that multiplies a variable.

2. Select the option that represents $a \times b$.

    ▪ a/b ▪ **ab** ▪ a.b

   1/3

3. $y + y + y$ can be written as ___.

   ▪ yyy ▪ y³ ▪ **3y**

   1/3

4. Select the correct notation for $a \times a$.

    ▪ aa ▪ **a²** ▪ 2a

   1/3

5. Select the option that represents $a \div b$.

    ▪ (i) ▪ (ii) ▪ **(iii)**

   1/3

   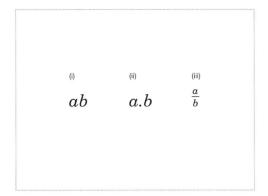

   $$\text{(i)} \quad ab \qquad \text{(ii)} \quad a.b \qquad \text{(iii)} \quad \frac{a}{b}$$

6. Select the option that represents $3 \times a$.

    ▪ **3a** ▪ a × a × a ▪ a3

   1/3

7. Select the correct notation for $(3x)^2$.

    ▪ 9x ▪ **9x²** ▪ 3x²

   1/3

8. Select the option that **does not** represent $\frac{1}{4} \times z$.

    ▪ **(i)** ▪ (ii) ▪ (iii)

   1/3

   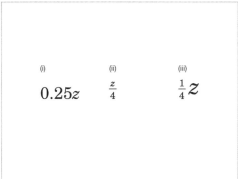

   $$\text{(i)} \quad 0.25z \qquad \text{(ii)} \quad \frac{z}{4} \qquad \text{(iii)} \quad \frac{1}{4}z$$

9. Select the option that represents $6 \div f$.

    ▪ **(i)** ▪ (ii) ▪ (iii)

   1/3

   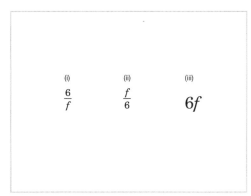

   $$\text{(i)} \quad \frac{6}{f} \qquad \text{(ii)} \quad \frac{f}{6} \qquad \text{(iii)} \quad 6f$$

10. Select the notation for $d \times 7$.

     ▪ d7 ▪ **7d** ▪ d × d × d × d × d × d × d

    1/3

---

**Level 2:** Fluency - More difficult notation.

✿ **Required:** 10/10　　✿ **Student Navigation:** on
✿ **Randomised:** off

---

11. Express the following as a single term:

    $f \times g$

    ▪ **fg**

12. What expression does $c^3$ represent?

     ▪ **c × c × c** ▪ c * c * c

---

Level 2: *cont.*

13. $x + x + x + x + x =$ __.

a b c ■ 5x

14. What is the following expression as a single term?

a b c ½ × $t$

■ t/2 ■ 1/2t

15. $15 \div e =$ __.

a b c ■ 15/e

16. How is $x \div y$ represented in algebraic notation?

a b c ■ x/y

17. $a \times 6 =$ __.

a b c ■ 6a

18. Give the following as a single term:

a b c $2g \times 3h$

■ 6gh

19. How is $r \times s \times t$ represented in algebraic notation?

a b c ■ rst

20. What expression does $m^2n$ represent?

a b c ■ m * m * n ■ m x m x n

# Expand Brackets

**Competency:** Use and interpret algebraic notation, including: brackets.

**Quick Search Ref:** 10183

Correct: Correct.    Wrong: Incorrect, try again.    Open: Thank you.

**Level 1:** Understanding - Multiply a bracket by a number.

✹ Required: 7/10    ✹ Student Navigation: on    ✹ Randomised: off

---

**1.** Multiply the bracket:

a
b
c

$5(4x + 3y)$.

▪ 20x + 15y  ▪ 15y + 20x

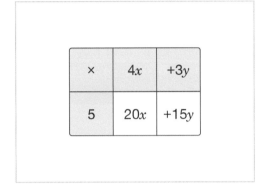

**2.** Expand the following:

a
b
c

$8(5m + 6)$.

▪ 40m + 48  ▪ 48 + 40m

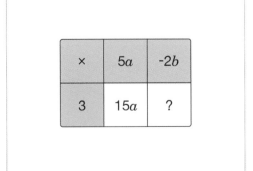

**3.** Multiply out $3(5a - 2b)$.

a
b
c

▪ -6b + 15a  ▪ 15a - 6b

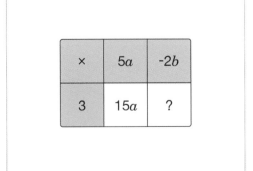

**4.** Expand the bracket:

a
b
c

$9(6z - 9)$.

▪ -81 + 54z  ▪ 54z - 81

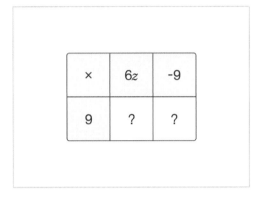

**5.** Expand $7(8r - 3)$.

a
b
c

▪ -21 + 56r  ▪ 56r - 21

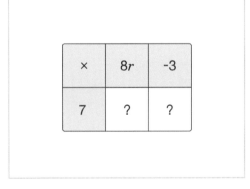

**6.** Multiply the bracket:

a
b
c

$-4(7c + 2d)$.

▪ -28c - 8d  ▪ -8d - 28c

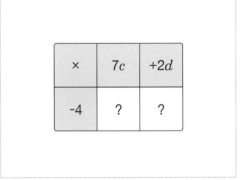

**7.** Expand the bracket:
a
b  3(9a + 7b).
c
   ▪ 21b + 27a ▪ 27a + 21b

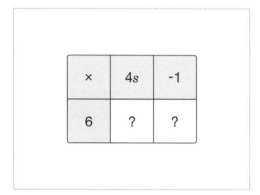

**8.** Multiply the bracket:
a
b  6(4s - 1).
c
   ▪ 24s - 6 ▪ -6 + 24s

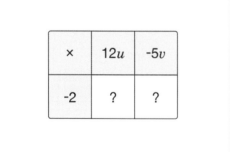

**9.** Expand the bracket:
a
b  -2(12u - 5v).
c
   ▪ 10v - 24u ▪ -24u + 10v

**10.** Multiply out the following:
a
b  12(7m - 3n).
c
   ▪ 84m - 36n ▪ -36n + 84m

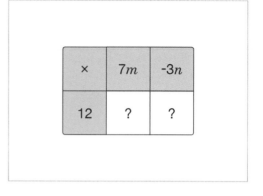

**Level 2:** Fluency - Multiply a bracket by a term.
✸ **Required:** 7/10  ✸ **Student Navigation:** on
✸ **Randomised:** off

**11.** Multiply the bracket:
a
b  3m(6n + 7).
c
   ▪ 18nm + 21m ▪ 21m + 18nm ▪ 21m + 18mn
   ▪ 18mn + 21m

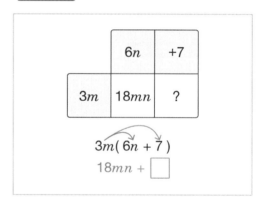

**12.** Expand the following expression:
a
b  6a(4b + 9c).
c
   ▪ 24ba + 54ca ▪ 24ab + 54ca ▪ 54ca + 24ab
   ▪ 24ab + 54ac ▪ 54ac + 24ab ▪ 54ca + 24ba
   ▪ 54ac + 24ba ▪ 24ba + 54ac

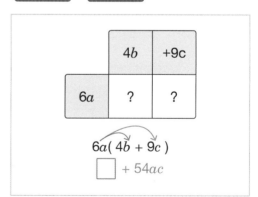

**13.** Multiply out $12x(5y - 3z)$.

**a**
**b**
**c**
- -36xz + 60xy
- -36zx + 60yx
- 60yx - 36zx
- 60yx - 36xz
- -36zx + 60yx
- 60xy - 36zx
- 60xy - 36xz
- -36xz + 60yx

**14.** Expand the bracket:
$7r(4r + 5)$.

- 28r + 35
- 28r² + 5r
- 28r² + 5
- 28r² + 35r

1/4

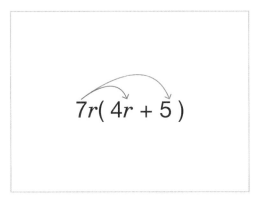

**15.** Multiply the bracket:
$4i(8j - k)$.

**a**
**b**
**c**
- -4ik + 32ij
- -4ik + 32ji
- 32ji - 4ki
- 32ji - 4ik
- -4ki + 32ji
- 32ij - 4ik
- 32ij - 4ki
- -4ki + 32ij

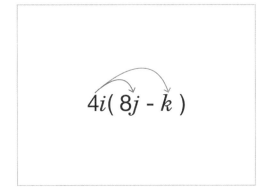

**16.** Expand $3r(9s - 2t)$.

**a**
**b**
**c**
- -6tr + 27rs
- 27rs - 6rt
- -6rt + 27sr
- 27rs - 6tr
- 27sr - 6tr
- 27sr - 6rt
- -6rt + 27rs
- -6tr + 27sr

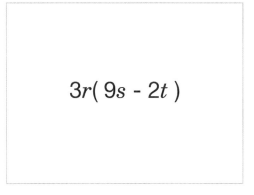

**17.** Multiply out $6a(5b + c - 8d)$.

- 30ab + 6ac - 48ad
- 30ab + 6ac + 48ad
- 30ab + 6ac - 8d
- 30ab + 6c - 48ad

1/4

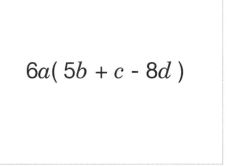

**18.** Expand the following:
$8m(7n + 5)$.

**a**
**b**
**c**
- 56nm + 40m
- 40m + 56nm
- 56mn + 40m
- 40m + 56mn

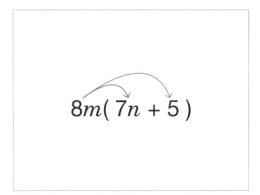

## Level 2: cont.

**19.** Multiply the bracket:

a b c $5r(5s - 9t)$.

- 25rs - 45rt
- 25sr - 45tr
- -45rt + 25rs
- 25rs - 45tr
- 25sr - 45tr
- -45tr + 25rs
- -45rt + 25sr
- -45rt + 25sr

$$5r( 5s - 9t )$$

**20.** Expand $4x(6x - 7y)$.

- 24x - 28y
- 24x² - 28xy
- 24x - 28xy
- 24x² + 28xy

1/4

$$4x( 6x - 7y )$$

## Level 3: Reasoning - Questions in context and comparing expressions.

✱ **Required:** 5/5  ✱ **Student Navigation:** on
✱ **Randomised:** off

**21.** Which two expressions expand to give the same answer?

2/5

- 6r(4s + 2t)
- 8r(3s + t)
- 3r(8s + 2t)
- 4r(6s + t)
- 2r(12s + 3t)

**22.** Write an expression without brackets for the area of the rectangle.

a b c

- 27a + 6ba
- 6ab + 27a
- 6ba + 27a
- 27a + 6ab

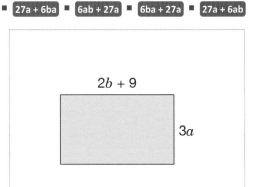

$2b + 9$

$3a$

**23.** Callum is $x$ years old and his sister is 2 years older than him. Their cousin is 3 times the age of Callum's sister. Write an expression without brackets for the age of Callum's cousin.

a b c

- 6 + 3x
- 3x + 6
- 3x + 6 years
- 6 + 3x years

**24.** Arrange the following expressions in ascending order (smallest value first).

↑ ↓

- 2(7 + 10t)
- 5(4t + 3)
- 4(5t + 4)

**25.** Yanis says that

a b c $3(2e + 5) + 4(2e + 5) = (3 + 4)(2e + 5)$.
Is he correct? Explain your answer.

## Level 4: Problem Solving - Multi-step problems and expanding and simplifying.

✱ **Required:** 5/5  ✱ **Student Navigation:** on
✱ **Randomised:** off

**26.** Form an expression for the area of the coloured shape shown.

a b c *Give your answer in its simplest form.*

- 10p + 6
- 6 + 10p

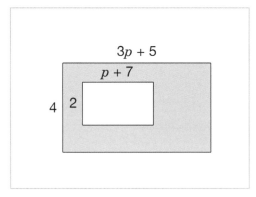

$3p + 5$

$p + 7$

$4$  $2$

**27.** Give an expression without brackets for the total surface area of the cuboid.

a b c *Give your answer in its simplest form.*

- 94 + 48x
- 48x + 94

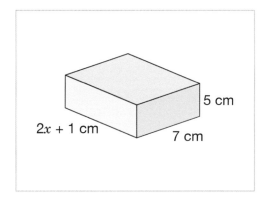

$2x + 1$ cm

5 cm

7 cm

**Level 4: cont.**

**28.** A square-based pyramid is made of triangles. Each
**a**  triangle has a base length of 6 metres and a
**b**  perpendicular height of $x + 7$ metres.
**c**  Write an expression for the total surface area of
the pyramid in square metres (m²).
*Don't include the units in your answer.*

▪ `12x + 120` ▪ `120 + 12x`

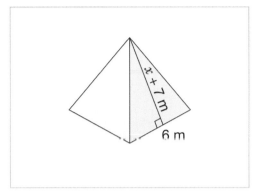

**29.** In an algebraic pyramid, each expression is the
**a**  sum of the two expressions directly below it. What
**b**  expression goes in the striped brick?
**c**

▪ `17x - 16` ▪ `-16 + 17x`

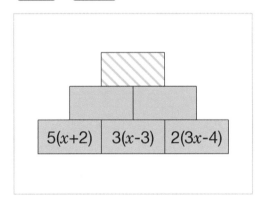

**30.** A square with sides $x + 5$ and a regular pentagon
**a**  with side length $2x - 3$ are joined as shown.
**b**  Write an expression without brackets for the
**c**  perimeter of the resulting shape.
*Simplify your answer.*

▪ `-5 + 12x` ▪ `12x - 5`

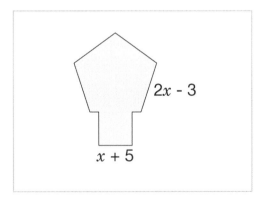

# Solve Linear 2-Step Equations

**Competency:** Use algebraic methods to solve linear equations in one variable (including all forms that require rearrangement).

**Quick Search Ref:** 10004

Correct: Correct.    Wrong: Incorrect, try again.    Open: Thank you.

**Level 1:** Understanding - Solve simple 1 and 2-step linear equations.

🌼 Required: 7/10    🌼 Student Navigation: on    🌼 Randomised: off

**1.** A mathematical statement showing that two expressions have equal value is:

- An expression - A term - A variable - [An equation]
1/5 - A coefficient

**2.** Which of the following is **not** an equation?

- x - 7 = 12 - 11 = 2y + 3 - [2a + 3b - 5ab]
- 3x + 2y = 2x - 15 - g = 2 + 3h
1/5

**3.** Which one of the following calculations can be used to solve the equation x - 3 = 15?

- 15 x 3 - 15 ÷ 3 - [15 + 3] - 15 - 3
1/4

**4.** Find the value of y in the following equation:

1
2
3
10 = y/5

- [50]

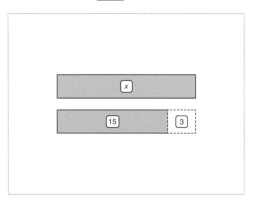

**5.** Solve the following equation to find the value of a:

1
2
3
2a + 3 = 7

- [2]

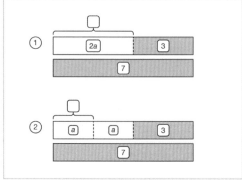

**6.** What is the value of t in the following equation?

1
2
3
19 = 3t + 4

- [5]

 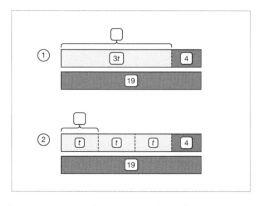

**7.** If 6y - 3 = 15, work out the value of y.

1
2
3

- [3]

 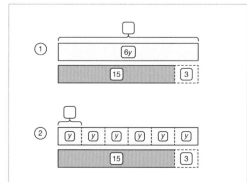

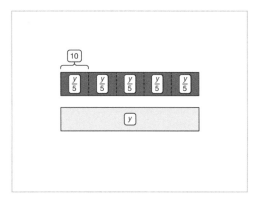

**8.** Solve the following equation to find the value of  *m*:

$5m + 2 = 27$

▪ **5**

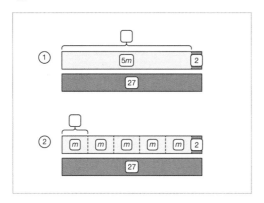

**9.** What is the value of *x* in the following equation?

$26 = 5 + 7x$

▪ **3**

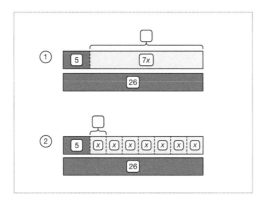

**10.** If $4c - 3 = 17$, work out the value of *c*.

▪ **5**

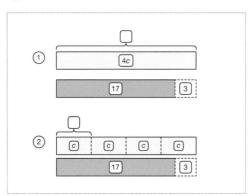

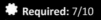 **Level 2:** Fluency - Solve 2-step linear equations with decimal and fractional answers.

✿ **Required:** 7/10     ✿ **Student Navigation:** on
✿ **Randomised:** off

**11.** Solve the following equation to find the value of *x*:

$x/2 + 5 = 12$

▪ **14**

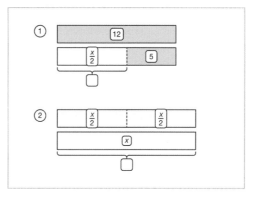

**12.** Solve the equation $x/4 - 2 = 1$ to find the value  of *x*.

▪ **12**

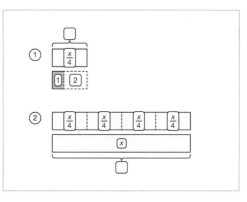

**13.** Which one of the four bar models can be used to represent the following equation?

$3x - 2 = 7$

1/4  ▪ **Bar model (a)**  ▪ **Bar model (b)**  ▪ **Bar model (c)**
▪ **Bar model (d)**

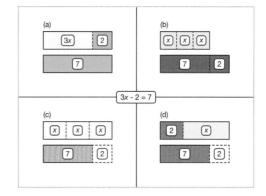

**Level 2:** *cont.*

**14.** What is the value of *a* in the following equation?
**1 2 3**  29 - 3*a* = 8

▪ **7**

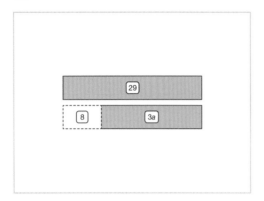

**15.** If 3y + 7 = 8, work out the value of *y*.
**a b c**  Give your answer as a fraction (using numbers not words).

▪ **1/3**

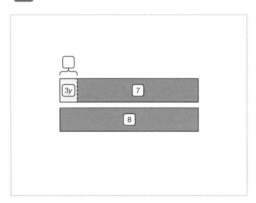

**16.** Find the value of *y* in the following equation:
**1 2 3**  23 = 8y + 19
Give your answer as a decimal.

▪ **0.5**

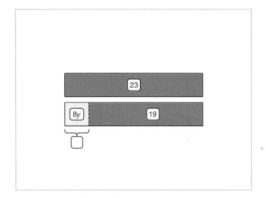

**17.** Solve the following equation to find the value of *b*:
**1 2 3**  3*b* + 1½ = 12
Give your answer as a decimal.

▪ **3.5**

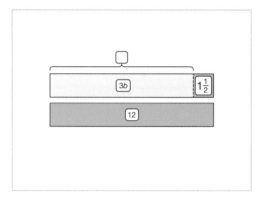

**18.** What is the value of *a* in the following equation?
**1 2 3**  6 = 10 - *a*/3

▪ **12**

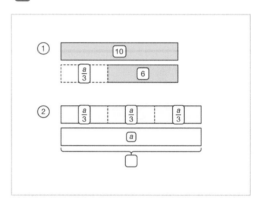

**19.** Solve the following equation to find the value of *x*:
**1 2 3**  5 + *x*/3 = 11

▪ **18**

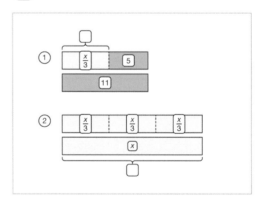

**20.** Find the value of *y* in the following equation:

**1**
**2**
**3**

4*y* - 1 = 5

Give your answer as a decimal.

▪ **1.5**

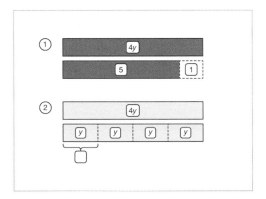

**Level 3:**   Reasoning - Solve equations in context.

✱ **Required:** 5/5   ✱ **Student Navigation:** on
✱ **Randomised:** off

**21.** Sort the following equations in **ascending** order according to the value of *x* in each equation.
You can make notes on the diagram to help you.

▪ Equation (c): 7 - 12x = 4   ▪ Equation (b): 16 = 9x - 29
▪ Equation (a): 8x - 51 = 5   ▪ Equation (d): 4 = 8 - x/3

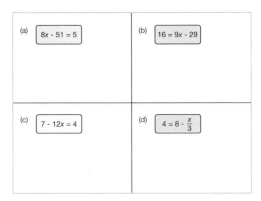

**22.** Identify **two** bar models that **cannot** be used to solve the equation 2*x* + 3 = 12?

**a**
**b**
**c**

Explain your answers.

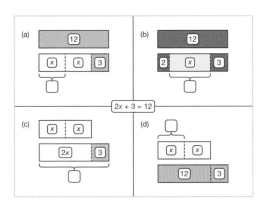

**23.** Here is Jacob's working for solving the equation ½*x* - 2 = 8.

**a**
**b**
**c**

1. What mistake has Jacob made?
2. What answer should he have got?
3. How could he have checked his answer?

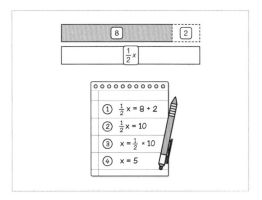

**24.** Madison thinks of a number.

**1**
**2**
**3**

She multiplies it by 3.
She adds 2 to the result.
The answer she gets is 29.
What number did Madison think of?

▪ **9**

**25.** Write an equation to represent the bar model shown and solve to find the value of *x*.

**a**
**b**
**c**

Can you write a **different** equation to represent the **same** bar model, and if so, will you get a different answer for *x*?
Explain your answer.

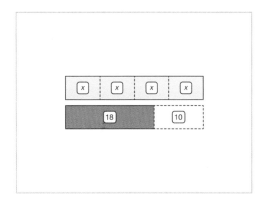

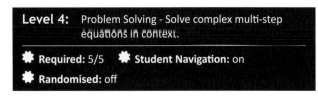

**Level 4:** Problem Solving - Solve complex multi-step equations in context.

---

⚹ **Required:** 5/5    ⚹ **Student Navigation:** on
⚹ **Randomised:** off

**26.** The diagram shows angles on a **straight line**.
Work out the value of $x$.

**1**
**2**
**3**    ▪ 19

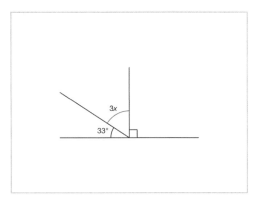

**27.** The diagram shows two **regular** polygons with the
**same** perimeter.
Work out the area of the square in cm²
(centimetres squared).
*Don't include units in your answer.*

a
b
c

▪ 36

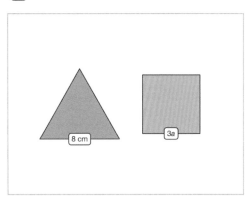

**28.** The diagram shows an incomplete bar model and
a function machine that both represent **the same**
equation.
Two **different** pieces of information are missing
from the diagram.
Work out the missing information to find the value
of $p$.

**1**
**2**
**3**

▪ 7

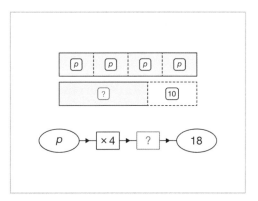

**29.** Each box in the pyramid is completed by **adding**
the terms in the **two boxes directly under it**.
The expression at the top of this equation pyramid
is equal to 26.
Complete the pyramid and use this information to
work out the value of $n$.

**1**
**2**
**3**

▪ 2.5

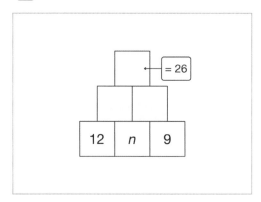

**30.** The shaded shape shown in the diagram has an
**area** of 34 cm².
Work out the value of $x$ in cm.
*Don't include units in your answer.*

**1**
**2**
**3**

▪ 12

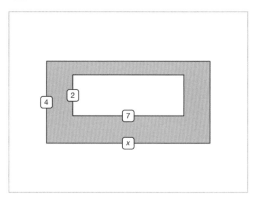

# Solve Linear 2-Step Equations (Balancing Method)

**Competency:** Use algebraic methods to solve linear equations in one variable (including all forms that require rearrangement).

**Quick Search Ref:** 10057

**Correct:** Correct.    **Wrong:** Incorrect, try again.    **Open:** Thank you.

---

**Level 1:** Understanding - Solve simple 2-step linear equations.

❋ **Required:** 7/10    ❋ **Student Navigation:** on    ❋ **Randomised:** off

---

**1.** Which option gives the correct method for solving the following equation?
$2y - 4 = 6$.

1/6

- Add 4 then multiply by 2.  ■ Subtract 4 then multiply by 2.
- Divide by 2 then add 4.  ■ Add 4 then divide by 2.
- Multiply by 2 then subtract 4.
- Subtract 4 then divide by 2.

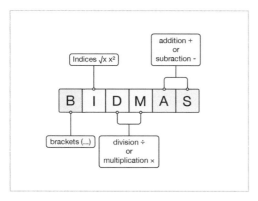

**2.** The shaded box shows a function machine for the equation $2x + 1 = 7$. Which diagram shows the correct **inverse function machine** for this equation?

1/4

- Inverse function machine A.
- Inverse function machine B.
- Inverse function machine C.
- Inverse function machine D.

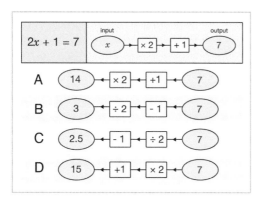

**3.** The equation $3x + 5 = 17$ has been solved using the balancing method. Select the two inverse operations which have been done on **both** sides of the equation to find the value of $x$.

1/4

- Add 5 then multiply by 3.  ■ Subtract 5 then divide by 3.
- Add 5 then divide by 3.  ■ Subtract 5 then multiply by 3.

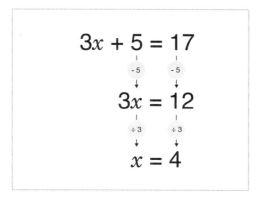

**4.** Which of the following is the correct first step for solving the equation $4a - 2 = 6$?

1/4

■ Solution A.  ■ Solution B.  ■ Solution C.  ■ Solution D.

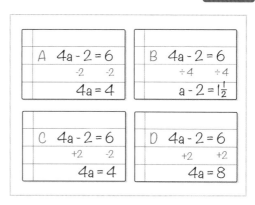

**5.** Solve the following equation to find the value of $t$:
$6t - 3 = 27$.

1
2
3    ■ 5

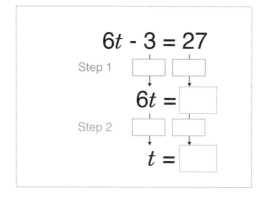

**Level 1:** *cont.*

**6.** If $3b + 11 = 35$, what is the value of $b$?

**1 2 3** ▪ **8**

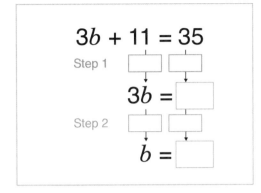

$$3b + 11 = 35$$

Step 1 ☐ ☐

$$3b = \boxed{\phantom{0}}$$

Step 2 ☐ ☐

$$b = \boxed{\phantom{0}}$$

**7.** Find the value of $y$ in the following equation: $8y - 7 = 25$.

**1 2 3** ▪ **4**

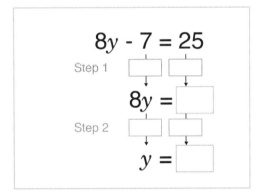

$$8y - 7 = 25$$

Step 1 ☐ ☐

$$8y = \boxed{\phantom{0}}$$

Step 2 ☐ ☐

$$y = \boxed{\phantom{0}}$$

**8.** Find the value of $a$ in the following equation: $5a - 9 = 21$.

**1 2 3** ▪ **6**

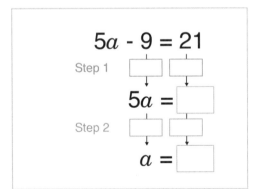

$$5a - 9 = 21$$

Step 1 ☐ ☐

$$5a = \boxed{\phantom{0}}$$

Step 2 ☐ ☐

$$a = \boxed{\phantom{0}}$$

**9.** If $8c + 7 = 31$, what is the value of $c$?

**1 2 3** ▪ **3**

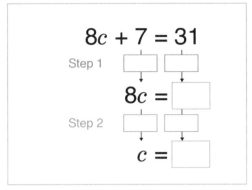

$$8c + 7 = 31$$

Step 1 ☐ ☐

$$8c = \boxed{\phantom{0}}$$

Step 2 ☐ ☐

$$c = \boxed{\phantom{0}}$$

**10.** Find the value of $x$ in the following equation: $6x + 5 = 47$.

**1 2 3** ▪ **7**

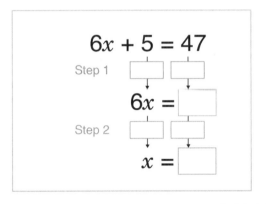

$$6x + 5 = 47$$

Step 1 ☐ ☐

$$6x = \boxed{\phantom{0}}$$

Step 2 ☐ ☐

$$x = \boxed{\phantom{0}}$$

**Level 2:** Fluency - Solve 2-step linear equations with negative, decimal and fractional answers.

✱ **Required:** 7/10     ✱ **Student Navigation:** on
✱ **Randomised:** off

**11.** Solve the following equation to find the value of $m$: $2m + 7 = 3$.

**1 2 3** ▪ **-2**

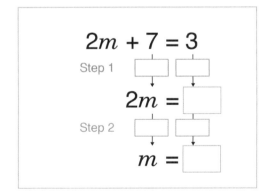

$$2m + 7 = 3$$

Step 1 ☐ ☐

$$2m = \boxed{\phantom{0}}$$

Step 2 ☐ ☐

$$m = \boxed{\phantom{0}}$$

**12.** Solve the following equation to find *x*:
5 = 6*x* + 2.
*Give your answer as a decimal.*

▪ 0.5

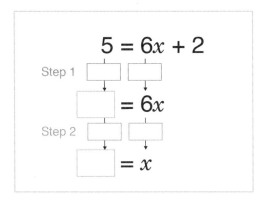

**13.** What is the value of *a* in the following equation?
21 - 3*a* = 6.

▪ 5

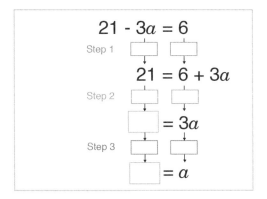

**14.** Work out the value of *n* in the following equation:
4*n* - 7 = -1.
*Give your answer as a decimal.*

▪ 1.5

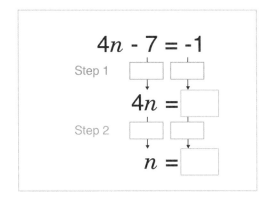

**15.** Work out the value of *x* in the equation shown in the diagram.

▪ 40

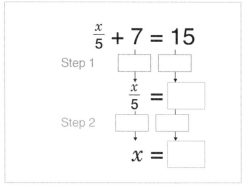

**16.** Solve the following equation to find the value of *n*:
-6 = 15 + 7*n*.

▪ -3

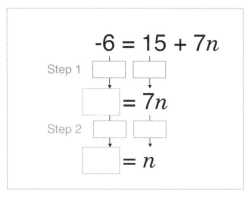

**17.** What is the value of *a* in the following equation?
11 - 9*a* = 8.
*Give your answer as a fraction in its simplest form.*

▪ 1/3

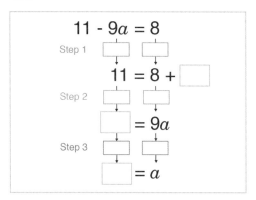

**Level 2: cont.**

**18.** Work out the value of *t* in the following equation:
$13 = 7 + 4t$.
*Give your answer as a decimal.*

1
2
3

▪ 1.5

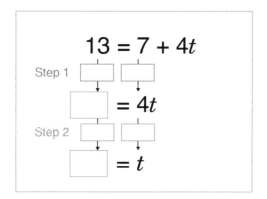

**19.** What is the value of *y* in the following equation?
$41 - 6y = 5$

1
2
3

▪ 6

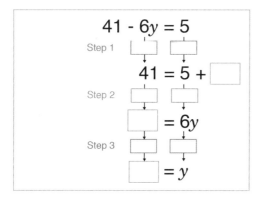

**20.** Work out the value of *x* in the equation shown in the diagram.

1
2
3

▪ 20

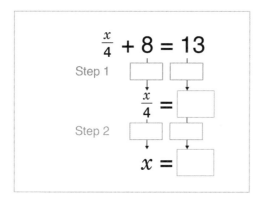

**Level 3:**   Reasoning - Solve equations in context.

✸ **Required:** 5/5   ✸ **Student Navigation:** on
✸ **Randomised:** off

**21.** The diagram shows a function machine.
Work out the input value, *x*.

1
2
3

▪ 8

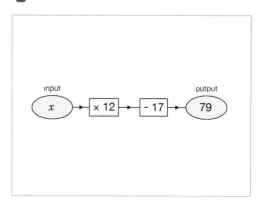

**22.** Sort the following equations in descending order (largest first) according to the value of *y* in each equation.

↑
↓

▪ Equation C ▪ Equation B ▪ Equation A
▪ Equation D

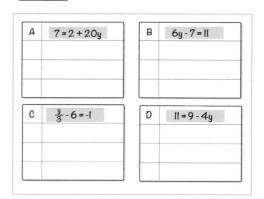

**23.** Alisha and Jack are discussing the solution to the following equation:
$6x - 42 = 0$.
Who do you agree with? Explain your answer.

a
b
c

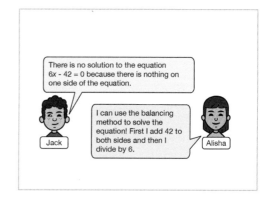

**24.** Rebecca has attempted to solve an equation as shown in the diagram. What mistake has Rebecca made? What answer should she have got?

a
b
c

$$\frac{x}{6} - 5 = 13$$
$$+5 \quad +5$$
$$\frac{x}{6} = 18$$
$$\times 6 \quad \times 6$$
$$6x = 108$$
$$\div 6 \quad \div 6$$
$$x = 18$$

**25.** The diagram shows how Manish solved an equation using the balancing method, but unfortunately, some of the ink got smudged. What was the equation that gave the solution $x = \frac{1}{2}$?

a
b
c

- 8 = 6x + 5  - 6x + 5 = 8  - 8 = 5 + 6x  - 5 + 6x = 8

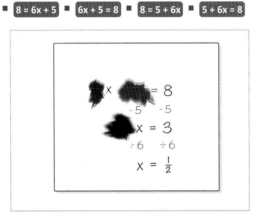

$$x \quad \blacksquare = 8$$
$$-5 \quad -5$$
$$x = 3$$
$$\div 6 \quad \div 6$$
$$x = \frac{1}{2}$$

**Level 4:** Problem Solving - Solve complex multi-step equations in context.

✹ **Required:** 5/5   ✹ **Student Navigation:** on
✹ **Randomised:** off

**26.** Each box in the pyramid is completed by adding the terms in the two boxes directly below it. If the expression at the top of this pyramid is equal to 13, what is the value of $x$?

1
2
3

- -5.5

*Pyramid with top box = 13, bottom row boxes: 7, $x$, 17*

**27.** Work out the value of $a$ if the perimeter of the regular hexagon is 14 centimetres (cm) longer than the perimeter of the rectangle.

1
2
3

- 11

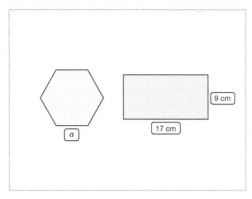

*Hexagon with side $a$; rectangle with sides 9 cm and 17 cm*

**28.** Fill in the blanks in the equation and the matching function machine and then solve the equation to find the value of $m$.

a
b
c

- 13

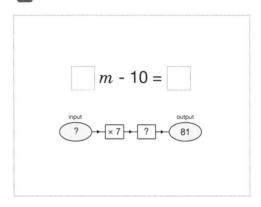

$$\boxed{\phantom{0}} \, m - 10 = \boxed{\phantom{0}}$$

input   →  ? →  × 7 →  ? →  output  81

**29.** Mr Moore gave the following instructions to his class:

1
2
3

'Think of a number, divide it by 5 and then subtract 3'.
If Ali's answer was 2, what number did he start with?

- 25

*Think of a number... divide it by 5... and then subtract 3...*

**30.** The diagram shows a function machine.

Work out the input value *x* that will give the same value for the output.

■

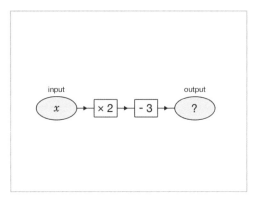

# Solve Linear Equations with Brackets

**Competency:** Use algebraic methods to solve simple linear equations in one variable where the variable appears on one side of the equation.

**Quick Search Ref:** 10164

**Correct:** Correct. **Wrong:** Incorrect, try again. **Open:** Thank you.

**Level 1:** Understanding - Expand brackets to simplify and solve equations.

✿ **Required:** 7/10 ✿ **Student Navigation:** on ✿ **Randomised:** off

---

**1.** How do you **expand** the expression 5(x + 4)?

1/6
- multiply the x in the brackets by 5
- add 5 to the 4 in the brackets
- multiply the 4 in the brackets by 5
- multiply both terms in the brackets by 5
- multiply the x in the brackets by 4
- add 5 to the terms in the brackets

**2.** What is the expanded for of 3(x + 5)?

1/5
- 3x + 5 ■ 3x + 8 ■ 3x + 15 ■ x + 15 ■ x³ + 15

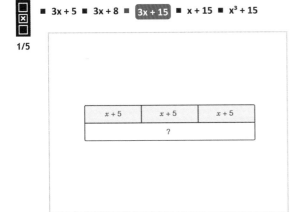

**3.** Expand 2(x + 8).

a b c
- 16 + 2x ■ 2x + 16

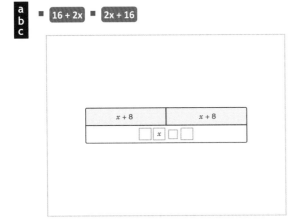

**4.** Multiply out the expression 12(y - 4).

a b c
- 12y - 48 ■ -48 + 12y

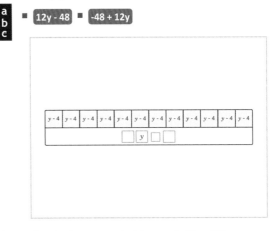

**5.** What is the expanded form of -5(x - 3)?

a b c
- 15 - 5x ■ -5x + 15

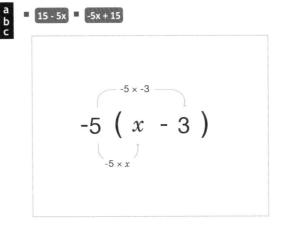

**6.** Expand and solve 3(x + 7) = 33.

1 2 3
- 4

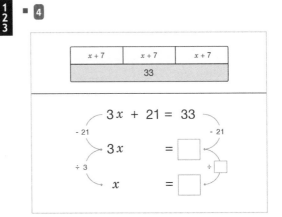

---

**7.** Find the solution of 8(k − 7) = -16

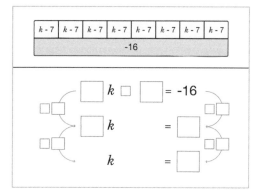

**8.** Find *x* when 9(*x* + 4) = 54.

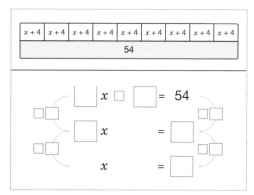

**9.** Solve 7(*x* − 2) = 21.

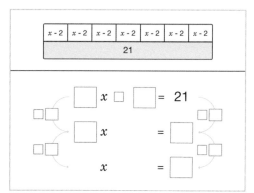

**10.** Give *y* when 4(*y* + 5) = 16.

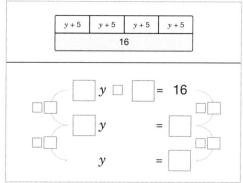

**Level 2:** Fluency - Solve equations with brackets, negatives and fractions

✷ **Required:** 7/10   ✷ **Student Navigation:** on
✷ **Randomised:** off

**11.** Solve 15 = 5(2 + *x*).

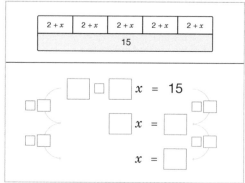

**12.** What value is needed to complete the multiplication grid?

■ -35

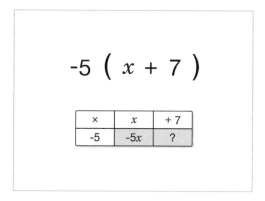

**13.** Find *y* when -49 = 7(*y* + 13).

■ -20

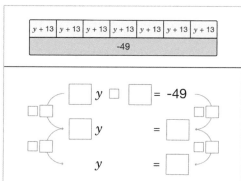

**14.** Find the value of $x$ when $6(10 - x) = 24$.

▪ **6**

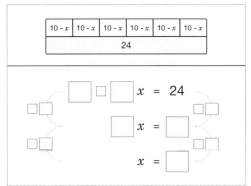

**15.** What is the solution for $4(t + \frac{1}{2}) = 14$.

▪ **3**

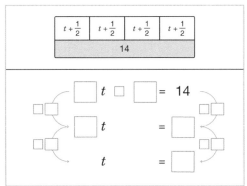

**16.** Solve $2(x + 4) = 7$.
*Give your answer as a decimal.*

▪ **-0.5**

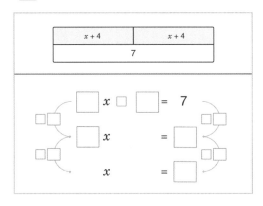

**17.** Give an equation (using brackets) for this function machine in its simplest form.

▪ **3(x + 5) = 24**

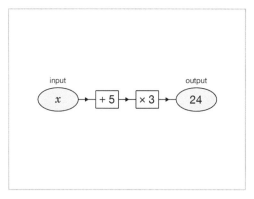

**18.** Solve $6(x + 3) = 21$.
*Give your answer as a decimal.*

▪ **0.5**

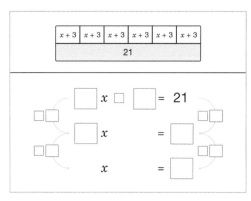

**19.** Give an equation (using brackets) for this function machine in its simplest form.
*Use decimals in your answer.*

▪ **0.5(x - 5) = 15**

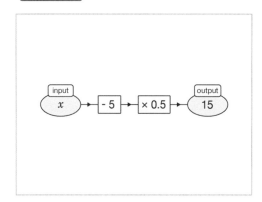

Level 2: *cont.*

**20.** What **term** is needed to complete the
a b c multiplication grid?

■ -3x

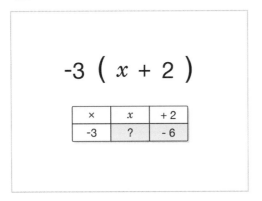

$$-3 \left( x + 2 \right)$$

| × | $x$ | + 2 |
|---|-----|-----|
| -3 | ? | - 6 |

**Level 3:** Reasoning - Forming and solving equations with brackets. Exploring and comparing methods.

✿ **Required:** 5/5    ✿ **Student Navigation:** on
✿ **Randomised:** off

**21.** Jane and Ali are discussing the equation $5(x - 8) =$
a b c 25. They each use different methods to solve the equation. Which do you think is the best method? Explain your answer.

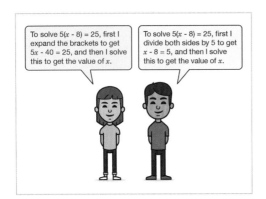

To solve $5(x - 8) = 25$, first I expand the brackets to get $5x - 40 = 25$, and then I solve this to get the value of $x$.

To solve $5(x - 8) = 25$, first I divide both sides by 5 to get $x - 8 = 5$, and then I solve this to get the value of $x$.

**22.** Aneesa is thinking of a number ($n$). She adds 17 to
1 2 3 it then multiplies the result by 5. The answer she gets is 100. What number did she start with?

■ 3

...add 17 first    ...then times by 5    ...equals 100

$n$

**23.** Monica has solved the equation $4(w + 4) = 28$. Is
a b c Monica coorrect? Explain your answer.

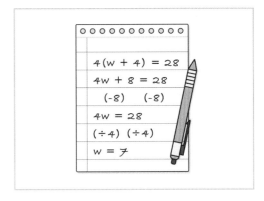

$4(w + 4) = 28$
$4w + 8 = 28$
$(-8) \quad (-8)$
$4w = 28$
$(\div 4) \quad (\div 4)$
$w = 7$

**24.** Find the value of $x$.

1 2 3   ■ 10

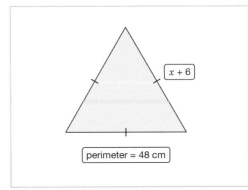

$x + 6$

perimeter = 48 cm

**25.** Sanjay has solved the equation $5(y - 2) = 20$. Is
a b c Sanjay correct? Explain your answer.

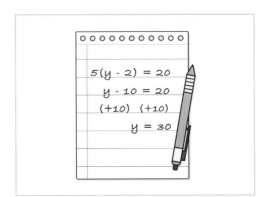

$5(y - 2) = 20$
$y - 10 = 20$
$(+10) \quad (+10)$
$y = 30$

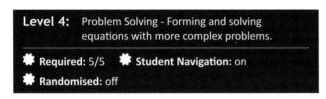

**Level 4:** Problem Solving - Forming and solving equations with more complex problems.

✱ **Required:** 5/5  ✱ **Student Navigation:** on
✱ **Randomised:** off

**26.** A farmer has 5 rows of apple trees with *x* number of trees in each row. If she plants 4 more apple trees in each row, there would be 60 trees in total. How many trees (*x*) were in each row to start with?

▪ 8

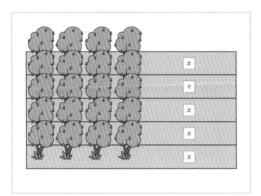

**27.** How old is Vijay (*v*)?

 ▪ 20

**28.** A rectangle has length (*x* + 5) cm and width 3 cm. The area of the rectangle is 51 cm². What is the **perimeter** of the rectangle?
*Include the units cm in your answer.*

▪ 40 cm  ▪ 40 centimetres

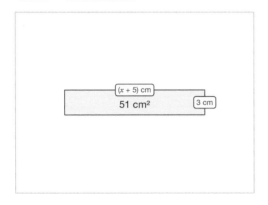

**29.** What is the area of rectangle (b) in cm²?
*Don't include the units in your answer.*

 ▪ 130

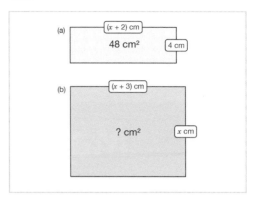

**30.** Ella is a ¼ of Tom's age (*t*). In 4 years time Jack will be twice as old as Tom. If Jack is 36 now, how old will Ella be in 4 years time?

▪ 8

# Mathematics Y7

## Measurement

### Area and Perimeter

# Calculate and Solve Problems Involving Perimeters of Composite Shapes

**Competency:** Calculate and solve problems involving: perimeters of 2D shapes (including circles), areas of circles and composite shapes.

**Quick Search Ref:** 10307

Correct: Correct.    Wrong: Incorrect, try again.    Open: Thank you.

**Level 1:** Understanding - Perimeter of composite shapes.

✱ Required: 10/10    ✱ Student Navigation: on    ✱ Randomised: off

1. Square A has an area of 64 square centimetres (cm²) and square B has an area of 25cm². What is the perimeter of the composite shape?
   *Include the units cm (centimetres) in your answer.*

   ▪ 72 centimetres ▪ 72 cm

   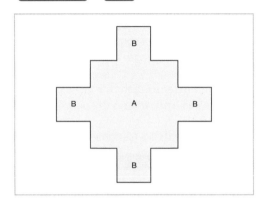

2. What is the perimeter of shape A?
   *Include the units cm (centimetres) in your answer.*

   ▪ 38 centimetres ▪ 38 cm

   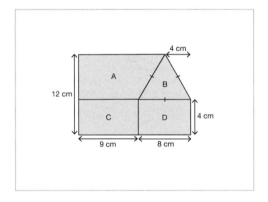

3. The square-shaped swimming pool has a width of 3 metres (m) and the lawn has an area of 45 square metres (m²).
   If the length and width of the garden both measure under 10 metres (m) and are whole numbers, what is the perimeter of the garden?
   *Include the units m (metres) in your answer.*

   ▪ 30 m ▪ 30 metres

   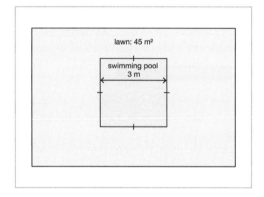

4. The playground at Hamilton School is made up of a square and two equilateral triangles. What is the perimeter of the playground?
   *Include the units m (metres) in your answer.*

   ▪ 54 m ▪ 54 metres

   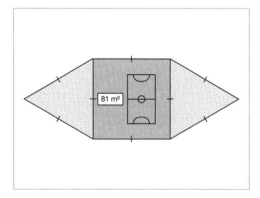

**Level 1: cont.**

**5.** A composite shape is made from an equilateral
**a** triangle and three identical squares. Calculate the
**b** perimeter of the composite shape in terms of x.
**c** *Give your answer in its simplest form.*

▪ 12x - 3

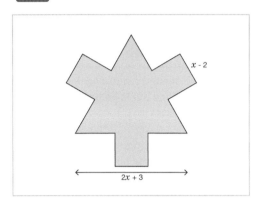

**6.** An equilateral triangle has an area of 18 square
**a** centimetres (cm²) and perpendicular height of 3
**b** centimetres (cm). What is its perimeter?
**c** *Include the units cm (centimetres) in your answer.*

▪ 36 centimetres  ▪ 36 cm

**7.** Darryl needs to buy new skirting boards for his
**a** master bedroom. How many metres (m) should he
**b** order?
**c** *Include the units m (metres) in your answer.*

▪ 30.8 metres  ▪ 30.8 m

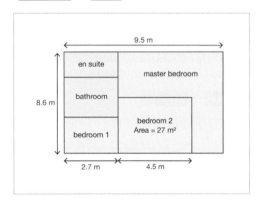

**8.** Lilly walks from the church (A) to the train station
**a** (B). How far does she walk in total?
**b** *Include the units m (metres) in your answer.*
**c**

▪ 36 m  ▪ 36 metres

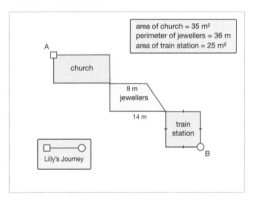

**9.** A composite shape is made from four identical
**a** rectangles and a larger rectangle. Calculate the
**b** perimeter of the composite shape in terms of x.
**c** *Give your answer in its simplest form.*

▪ 14x + 26

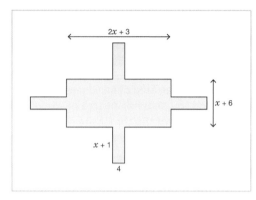

**10.** The vertices of the rhombus align with the
**a** midpoints in the length and width of the
**b** rectangle.
**c** If the perimeter of each triangle is 21.2
centimetres (cm), what is the perimeter of the
rhombus?
*Include the units cm (centimetres) in your answer.*

▪ 35.2 cm  ▪ 35.2 centimetres

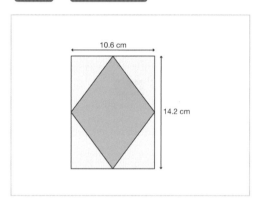

# Calculate and Solve Problems Involving the Area of Composite Shapes

**Competency:** Calculate and solve problems involving: perimeters of 2D shapes (including circles), areas of circles and composite shapes.

**Quick Search Ref:** 10252

**Correct:** Correct.    **Wrong:** Incorrect, try again.    **Open:** Thank you.

**Level 1:** Problem Solving - Calculating the area of compound shapes.

✸ **Required:** 10/10        ✸ **Student Navigation:** on        ✸ **Randomised:** off

---

**1.** Harry's grandad has a square vegetable patch and nine square flower beds in his garden. In square metres (m²), what is the area of the lawn where he grows apples?
*Don't include the units in your answer.*

▪ 27

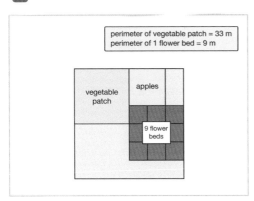

**2.** The area of the shaded cuboid face is _____cm²
Calculate the area of the shaded cuboid face in square centimetres (cm²).
*Don't include the units in your answer.*

▪ 90

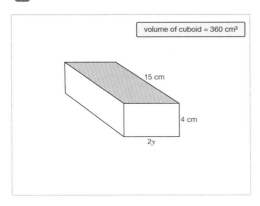

**3.** A family is tiling their kitchen and utility room floors. If tiles cost £12.50 per square metre (m²), how much will the tiles cost in total?
*Include the £ sign in your answer.*

▪ £418.75

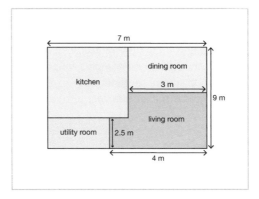

**4.** Riley is making a flag of the Bahamas for a parade. What is the minimum amount of yellow material he needs?
*Don't include the units in your answer.*

▪ 68

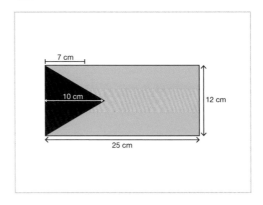

**5.** Calculate the purple shaded area of the right-angled trapezium.
*Don't include the units in your answer.*

1
2
3

▪ 743

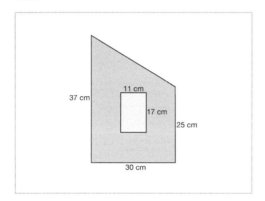

**6.** Triangle A has a perimeter of 57 centimetres (cm) and triangle B has a perimeter of 39 centimetres. If both are equilateral triangles, what is the area of shape C in square centimetres (cm²)?
*Don't include the units in your answer.*

1
2
3

▪ 247

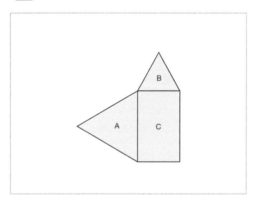

**7.** If turf costs £3.50 per square metre (m²), how much will it cost to returf Jamie's lawn?
*Include the £ sign in your answer.*

a
b
c

▪ £176.75

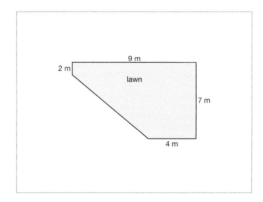

**8.** Some friends are making a Trinidad and Tobago flag for a school project. The ratio of black:white is 3:2. How much white card do they need in square cm (cm²)?
*Don't include units in your answer.*

1
2
3

▪ 150

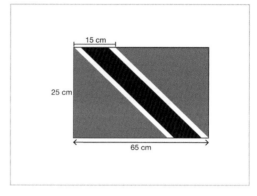

**9.** The arrowhead is made from a rectangle and an equilateral triangle. What is the area of the arrowhead in terms of $x$?
*Give your answer in its simplest form.*

a
b
c

▪ 13x + 6   ▪ 6 + 13x

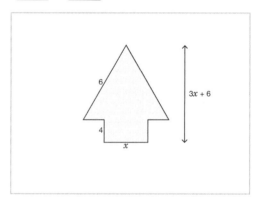

**10.** The area of the large square is 169 cm². The three smaller squares each have an area of $x^2$. The areas where the smaller squares overlap measure 1 cm². What is the value of $x$?

1
2
3

▪ 5

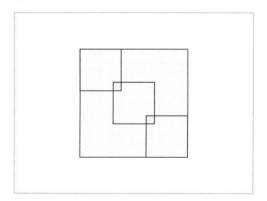

# Mathematics

**Y7**

## Geometry

2D Shape

Angles

Transformation

# Describe and use conventional terms and notations of 2D shapes

**Competency:** Describe, sketch and draw using conventional terms and notations: points, lines, parallel lines, perpendicular lines, right angles, regular polygons, and other polygons that are reflectively and rotationally symmetric.

**Quick Search Ref:** 10054

**Correct:** Correct.    **Wrong:** Incorrect, try again.    **Open:** Thank you.

**Level 1:** Understanding - Conventional terms and notations of 2D shapes.

⬥ **Required:** 10/10          ⬥ **Student Navigation:** on          ⬥ **Randomised:** off

---

**1.** Which line measures 7 centimetres (cm)?

a
b
c

▪ Line BC  ▪ CB  ▪ Line CB  ▪ BC

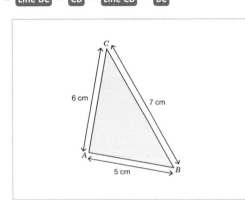

**2.** Angle *BCD* can also be know as:

▪ Angle v  ▪ Angle w  ▪ Angle x  ▪ Angle y  ▪ Angle z

1/5

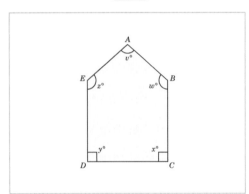

**3.** Which diagram shows a pair of parallel lines?

 ▪ (i)  ▪ (ii)  ▪ (iii)  ▪ (iv)

1/4

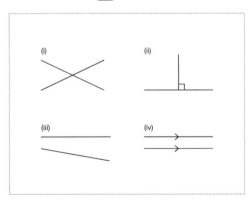

**4.** Which two parallelograms are labelled correctly?

▪ (i)  ▪ (ii)  ▪ (iii)  ▪ (iv)  ▪ (v)

2/5

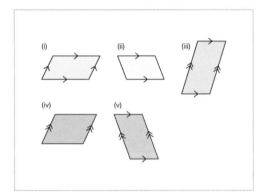

**5.** What is the size of angle *ABC*?
*Don't include the units in your answer.*

1
2
3

▪ 90

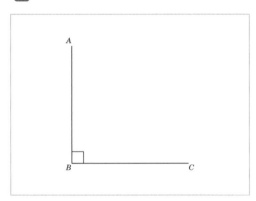

**6.** Select the diagram which shows that line *AB* is perpendicular to line *CD*.

 ▪ (i)  ▪ (ii)  ▪ (iii)  ▪ (iv)

1/4

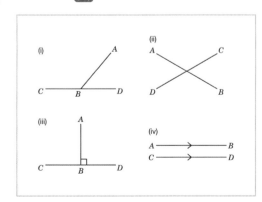

---

**7.** What is the length of line *AB*?
*Include the units in your answer.*

a
b
c   ▪ 7 cm  ▪ 7 centimetres

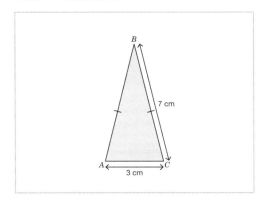

**8.** Which line has the same length as line *CD*?

a
b
c   ▪ Line BA  ▪ Line AB  ▪ BA  ▪ AB

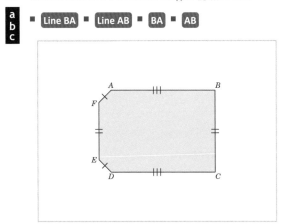

**9.** Which two oblongs are labelled correctly?

  ▪ (i)  ▪ (ii)  ▪ (iii)  ▪ (iv)  ▪ (v)

2/5

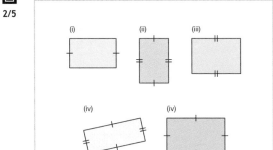

**10.** Which angle is the same size as angle *DAB*?

a
b
c   ▪ DCB  ▪ BCD  ▪ Angle BCD  ▪ Angle DCB

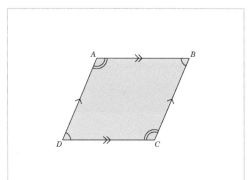

## Identify all parts of a circle

**Competency:** Derive and illustrate properties of triangles, quadrilaterals, circles, and other plane figures [for example, equal lengths and angles] using appropriate language and technologies.

**Quick Search Ref:** 10047

**Correct:** Correct.    **Wrong:** Incorrect, try again.    **Open:** Thank you.

**Level 1:** Understanding - Identify all parts of a circle (including origin, segment, sector, chord, arc and tangent).

✱ **Required:** 10/10    ✱ **Student Navigation:** on    ✱ **Randomised:** off

1. Circumference is:

1/4
- the distance from the origin of the circle to any point on its circumference.
- **the distance around the edge of a circle.**
- a straight line which passes through the origin of the circle connecting two points on the circumference.
- the total surface of a circle.

2. Which part shows the **origin** of the circle?

1/4
■ A ■ B ■ **C** ■ D

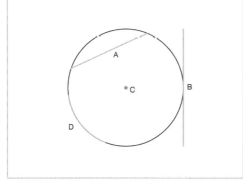

3. The **arc** of a circle is:

1/4
- **a curved line which is part of the circumference.**
- the distance around the edge of a circle.
- a straight line which passes through the origin of the circle connecting two points on the circumference.
- the distance from the origin of the circle to any point on its circumference.

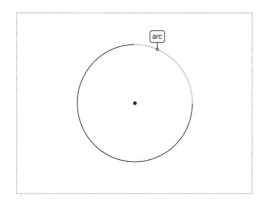

4. What does the **radius** of the circle measure?
*Include the units cm (centimetres) in your answer.*

- **7 cm** ■ **7 centimetres**

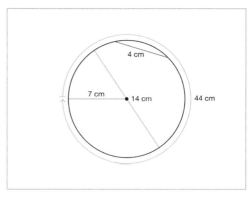

5. Select the **tangent**.

1/4
■ A ■ **B** ■ C ■ D

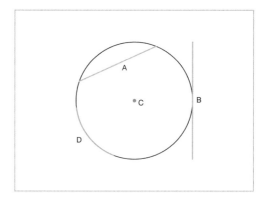

6. The **chord** measures ___ centimetres (cm).

- **18**

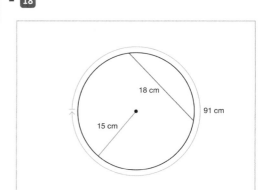

**7.** What does the **diameter** of the circle measure?

abc *Include cm (centimetres) in your answer.*

▪ 14 centimetres  ▪ 14 cm

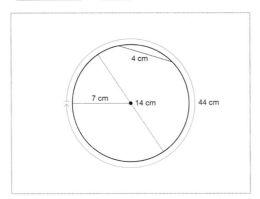

**8.** What is the area of the **segment** in square

123 centimetres (cm²)?
*Don't include the units in your answer.*

▪ 52

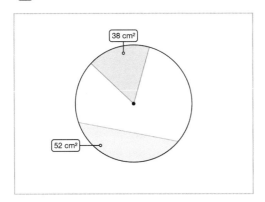

**9.** What is the area of the **sector** in square

123 centimetres (cm²)?
*Don't include the units in your answer.*

▪ 38

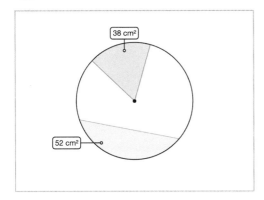

**10.** What fraction of a circle is a quadrant?

abc *Give your answer in its simplest form.*

▪ 1/4

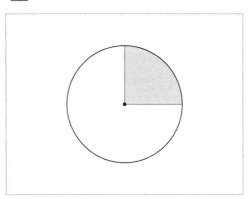

# Identify Reflective Symmetry of Polygons

**Competency:** Describe, sketch and draw using conventional terms and notations: points, lines, parallel lines, perpendicular lines, right angles, regular polygons, and other polygons that are reflectively and rotationally symmetrical.

**Quick Search Ref:** 10326

Correct: Correct.    Wrong: Incorrect, try again.    Open: Thank you.

**Level 1:** Understanding - Identifying reflective symmetry.

✿ **Required:** 6/8    ✿ **Student Navigation:** on    ✿ **Randomised:** off

1. Which statement describes reflective symmetry?

1/3

- When a shape still looks the same after a rotation between 0° and 360° around its centre.
- The number of positions a shape can be rotated to between 0° and 360° around its centre and still look the same.
- When an imaginary line divides a shape into two mirror image halves.

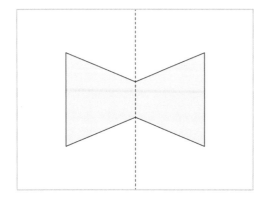

2. Which two images show reflective symmetry?

2/4

■ (i) ■ (ii) ■ (iii) ■ (iv)

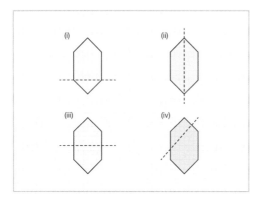

3. Which shape has only two lines of reflective symmetry?

1/4

■ (i) ■ (ii) ■ (iii) ■ (iv)

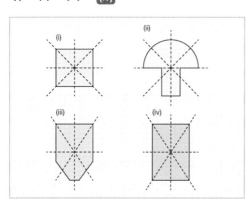

4. Which two shapes have only one line of reflective symmetry?

2/4

■ (i) ■ (ii) ■ (iii) ■ (iv)

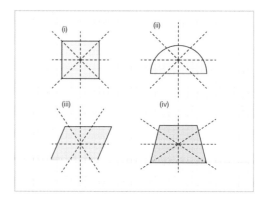

5. Which shape has four lines of reflective symmetry?

1/4

■ (i) ■ (ii) ■ (iii) ■ (iv)

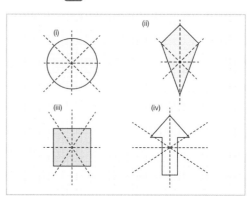

6. Which image shows reflective symmetry?

1/4

■ (i) ■ (ii) ■ (iii) ■ (iv)

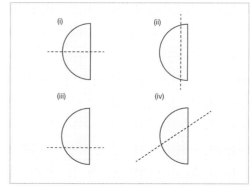

**7.** Which shape has only two lines of reflective symmetry?

 ▪ (i) ▪ (ii) ▪ (iii) ▪ (iv)

1/4

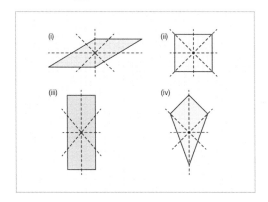

**8.** Reflective symmetry is when an imaginary line divides a shape into two mirror image halves.

 ▪ True ▪ False

1/2

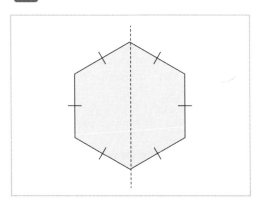

---

**Level 2:** Fluency - Identifying total lines of symmetry.

❋ **Required:** 6/8     ❋ **Student Navigation:** on
❋ **Randomised:** off

---

**9.** How many lines of reflective symmetry does the shape have?

 ▪ 0

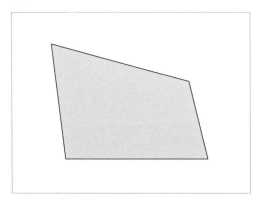

**10.** How many lines of reflective symmetry does the shape have?

 ▪ 1

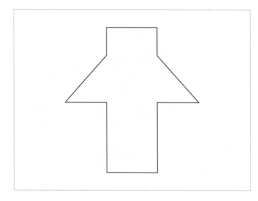

**11.** Which **two** polygons have five or more lines of reflective symmetry?

 ▪ (i) ▪ (ii) ▪ (iii) ▪ (iv)

2/4

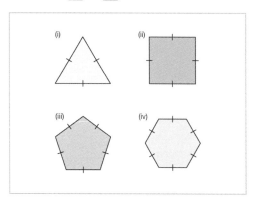

**12.** How many lines of symmetry does a parallelogram have?

 ▪ 0

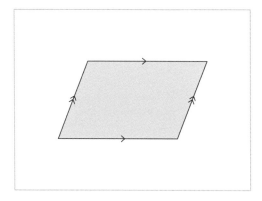

Level 2: *cont.*

**13.** How many lines of symmetry does a circle have?

 ■ 2 ■ 4 ■ 360 ■ infinite

1/4

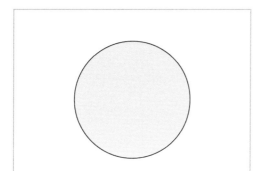

**14.** Select the two shapes that have no lines of reflective symmetry.

 ■ (i) ■ (ii) ■ (iii) ■ (iv)

2/4

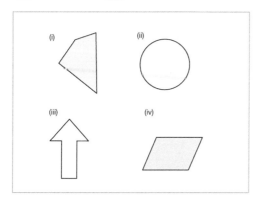

**15.** How many lines of reflective symmetry does a rhombus have?

 ■ 2

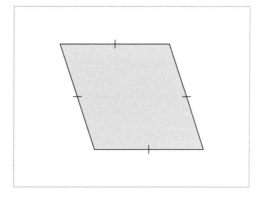

**16.** How many lines of reflective symmetry does this shape have?

 ■ 2

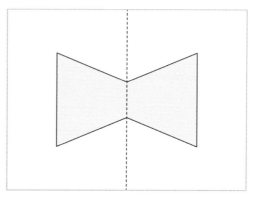

**Level 3:**   Reasoning - With reflective symmetry.

✿ **Required:** 3/3    ✿ **Student Navigation:** on

✿ **Randomised:** off

**17.** Christian says, "A square has four sides and four lines of symmetry. So, all quadrilaterals must have the same number of lines of symmetry as the number of sides". Is Christian correct? Explain your answer.

**18.** How many lines of symmetry are there in a regular *n*-sided shape?
Give your answer in terms of *n*.

■ n

**19.** Jade says, "Out of the four shapes shown the dodecagon must have the most lines of symmetry because it has the greatest number of sides. Is Jade correct? Explain your answer.

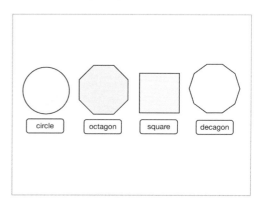

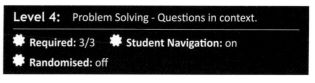

**Level 4:**   Problem Solving - Questions in context.

✿ **Required:** 3/3    ✿ **Student Navigation:** on

✿ **Randomised:** off

**20.** Gina has 50 square pieces of paper. What is the greatest number of squares she can use to make a shape with four lines of reflective symmetry?

■ 49

**21.** How many **more** parts of the triangle need to be shaded to give the triangle three lines of symmetry?

▪ 13

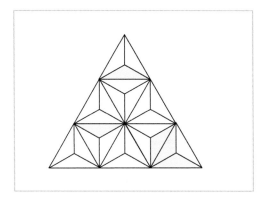

**22.** How many ways can you arrange the tiles in the grid to make a shape with reflective symmetry?

▪ 14

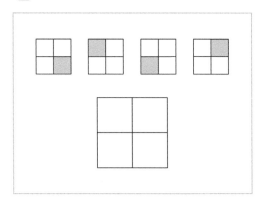

# Identify Rotational Symmetry of Polygons

**Competency:** Describe, sketch and draw using conventional terms and notations; points, lines, parallel lines, perpendicular lines, right angles, regular polygons, and other polygons that are reflectively and rotationally symmetrical.

**Quick Search Ref:** 10325

Correct: Correct.    Wrong: Incorrect, try again.    Open: Thank you.

**Level 1:** Understanding - Identifying rotational symmetry.

✿ **Required:** 6/8    ✿ **Student Navigation:** on    ✿ **Randomised:** off

**1.** Select the statement that describes rotational symmetry.

1/3

- The number of positions a shape can be rotated to between 0° to 360° around its centre and still look the same.
- When an imaginary line divides a shape into two mirror image halves.
- When a shape still looks the same after a rotation between 0° and 360°.

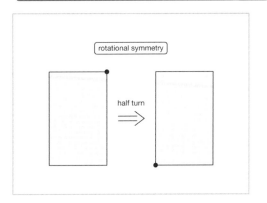

**2.** Which two examples show rotational symmetry of shape A?

2/4

■ (i)  ■ (ii)  ■ (iii)  ■ (iv)

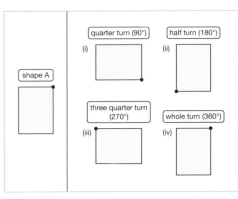

**3.** A rectangle has rotational symmetry of order __.

1/4

■ 1  ■ 2  ■ 3  ■ 4

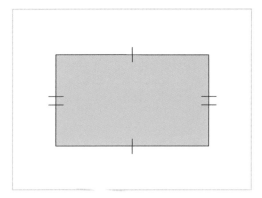

**4.** An equilateral triangle has rotational symmetry of order __.

1/5

■ 2  ■ 3  ■ 4  ■ 5  ■ 6

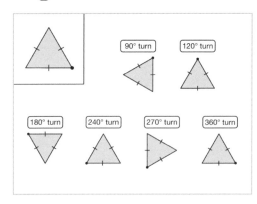

**5.** A square has rotational symmetry of order __.

1
2
3

■ 4

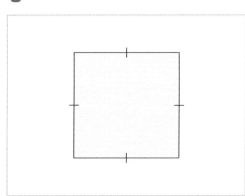

**Level 1: cont.**

**6.** An isosceles triangle has rotational symmetry of order __.

1
2
3  ▪ 1

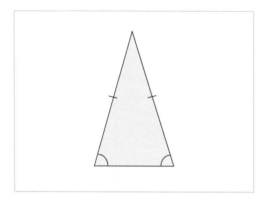

**7.** Which two examples show rotational symmetry of shape A?

☐
☒  ▪ (i) ▪ (ii) ▪ (iii) ▪ (iv)
☐

2/4

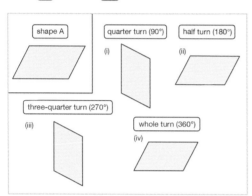

**8.** A scalene triangle has rotational symmetry of order __.

1
2
3  ▪ 1

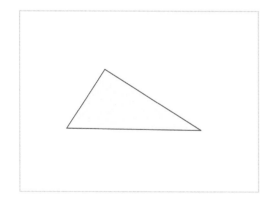

**Level 2:**   Fluency - Identifying order of rotational symmetry.

✿ **Required:** 6/8   ✿ **Student Navigation:** on
✿ **Randomised:** off

**9.** What is the order of rotational symmetry of a regular pentagon?

1
2
3  ▪ 5

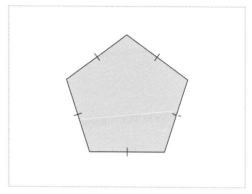

**10.** Select the polygon that has the same order of rotational symmetry as a rhombus.

☐
☒  ▪ (i) ▪ (ii) ▪ (iii) ▪ (iv)
☐

1/4

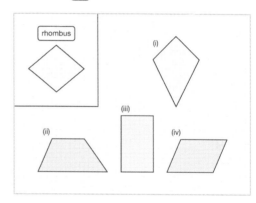

**11.** What is the order of rotational symmetry of the dodecagon?

1
2
3  ▪ 4

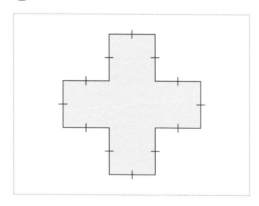

**12.** What is the order of rotational symmetry of a regular nonagon?

 ■ 9

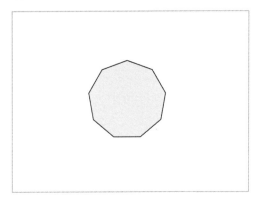

**13.** Select the polygon that has the same order of rotational symmetry as a kite.

 ■ (i) ■ (ii) ■ (iii) ■ (iv)

1/4

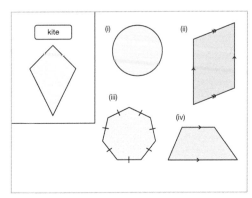

**14.** What is the order of rotational symmetry of the octagon?

 ■ 2

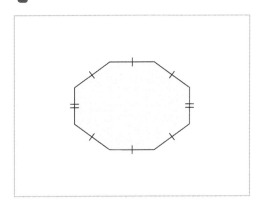

**15.** Select the polygon which has the same order of rotational symmetry as a rectangle.

 ■ (i) ■ (ii) ■ (iii) ■ (iv)

1/4

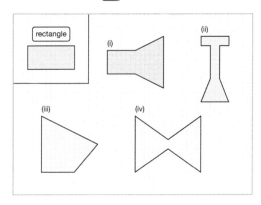

**16.** What is the order of rotational symmetry of the irregular hexagon?

 ■ 1

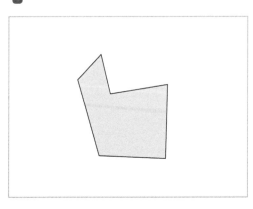

**Level 3:**    Reasoning - With rotational symmetry.

✱ **Required: 3/3**    ✱ **Student Navigation: on**
✱ **Randomised: off**

**17.** What is the smallest number of degrees you need to rotate a regular pentagon by to get rotational symmetry?
*Don't include the units in your answer.*

■ 72

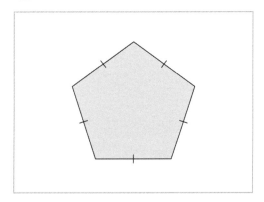

**18.** Tony says, "A square has rotational symmetry of order 4 and four lines of symmetry. So, all quadrilaterals must have the same order of rotational symmetry as lines of symmetry".
Is Tony correct? Explain your answer.

**Level 3:** *cont.*

19. Nathan says, "The octagon has rotational
**a** symmetry of order 8".
**b** Is Nathan correct? Explain your answer.
**c**

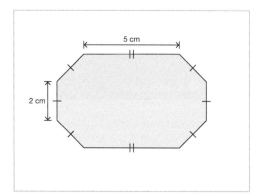

**Level 4:**   Problem Solving - Rotational symmetry in context.

✹ **Required:** 3/3   ✹ **Student Navigation:** on
✹ **Randomised:** off

20. How many different ways can the jigsaw piece be
**1** placed in the board?
**2**
**3**   ▪ **4**

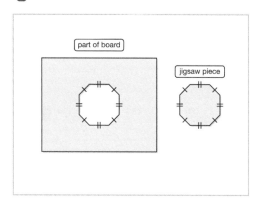

21. If two cogs start to rotate and then stop the next
**1** time A and B touch, how many full rotations will
**2** cog A make before the cogs come to a stop?
**3** *Both cogs are regular polygons.*

   ▪ **3**

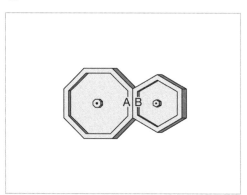

22. What is the minimum number of squares Jenny
**1** needs to shade to give the shape rotational
**2** symmetry of order 2?
**3**

   ▪ **5**

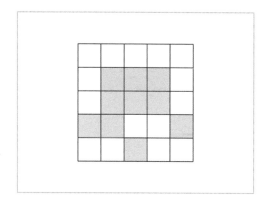

# Apply the properties of angles at a point, angles on a straight line and vertically opposite angles

**Competency:** Apply the properties of angles at a point, angles at a point on a straight line, vertically opposite angles.

**Quick Search Ref:** 10273

Correct: Correct.    Wrong: Incorrect, try again.    Open: Thank you.

**Level 1:** Applying the properties of angles at a point, on a straight line and vertically opposite.

✿ **Required:** 10/10    ✿ **Student Navigation:** on    ✿ **Randomised:** off

---

1. There are seven angles at a point. One angle measures 48 degrees and the remaining angles are equal. What is their size?
   *Don't include units in your answer.*

   ▪ 52

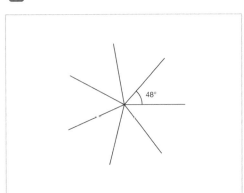

2. The diagram shows a regular pentagon. What is the size of angle *x*?
   *Don't include units in your answer.*

   ▪ 72

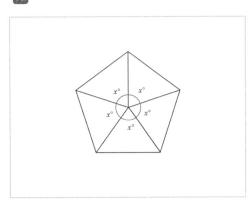

3. Billy was given four angles to draw around a point but made a mistake with his measurements. Three of the angles are 105°, 54° and 20°. How many obtuse angles should his diagram show?

   1/5

   ▪ 1 ▪ 2 ▪ 3 ▪ 4 ▪ 0

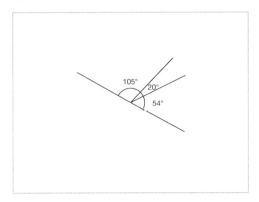

4. Select the equation that represents the size of angle *y*.

   1/4

   ▪ y + z = 180 ▪ y = z - 180 ▪ y = 360 - z ▪ y = 180 - z

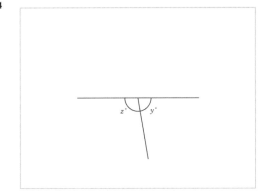

---

**5.** If *b* + *c* = 130, what is the value of *a*?
*Don't include units in your answer.*

■ 115

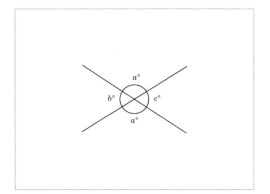

**6.** Calculate the value of *z*.

■ 80

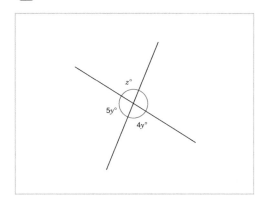

**7.** If *e* + *f* = 104, what is the value of *d*?
*Don't include units in your answer.*

■ 128

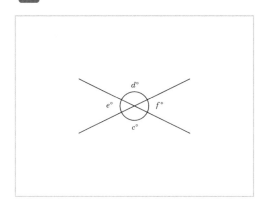

**8.** Calculate the value of *k*.
*Don't include units in your answer.*

■ 30

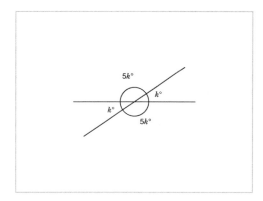

**9.** Calculate the size of angle *c*.
*Don't include units in your answer.*

■ 106

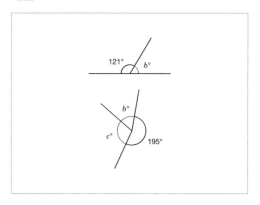

**10.** Angles *x* and *y* are drawn on a straight line. Angle *y* is 4 times bigger than angle *x*.
What is the size of angle *y*?
*Don't include units in your answer.*

■ 144

# Derive and use the sum of angles in a triangle and a quadrilateral

**Competency:** Derive and use the sum of angles in a triangle and use it to deduce the angle sum in any polygon, and to derive properties of regular polygons and a quadrilateral.

**Quick Search Ref:** 10215

**Correct:** Correct.     **Wrong:** Incorrect, try again.     **Open:** Thank you.

**Level 1:** Understanding - Use properties of angles in triangles and quadrilaterals to find missing angles.

✿ **Required:** 7/10          ✿ **Student Navigation:** on          ✿ **Randomised:** off

**1.** What is the sum of all the angles in a quadrilateral?
*Don't include the units in your answer.*

▪ 360

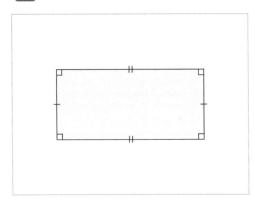

**2.** What is the sum of all the angles in a triangle?
*Don't include the units in your answer.*

▪ 180

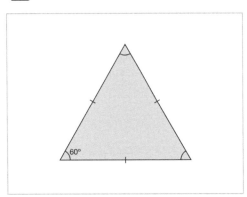

**3.** What type of triangle is shown?

▪ Scalene ▪ Isosceles ▪ Equilateral ▪ Right angle

1/4

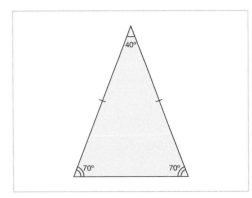

**4.** Calculate the sum of *a* + *b*.
*Don't include the units in your answer.*

▪ 90

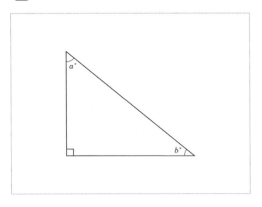

**5.** Calculate the size of angle *ABC*.
*Don't include the units in your answer.*

▪ 75

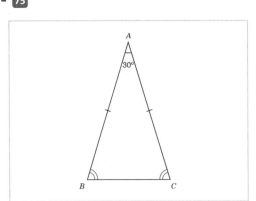

**6.** Calculate the size of angle *BCD*.
*Don't include the units in your answer.*

▪ 125

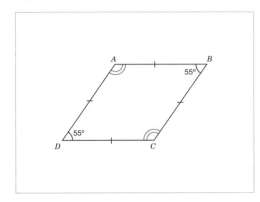

**Level 1:** *cont.*

**7.** What is the size of angle *x*?
*Don't include the units in your answer.*

1
2
3 ▪ 137

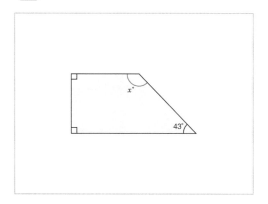

**8.** Calculate the sum of *y* + *z*.
*Don't include the units in your answer.*

1
2
3 ▪ 60

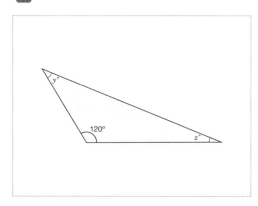

**9.** Calculate the size of angle *XYZ*.
*Don't include the units in your answer.*

1
2
3 ▪ 135

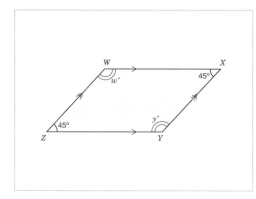

**10.** Calculate the size of angle *DEF*.
*Don't include the units in your answer.*

1
2
3 ▪ 55

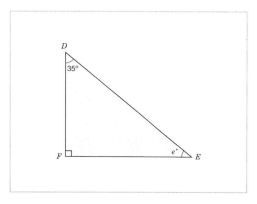

**Level 2:** Fluency - Find missing angles in triangles and quadrilaterals (algebra).

✱ **Required:** 7/10   ✱ **Student Navigation:** on
✱ **Randomised:** off

**11.** Calculate the size of angle *a*.
*Don't include the units in your answer.*

1
2
3 ▪ 66

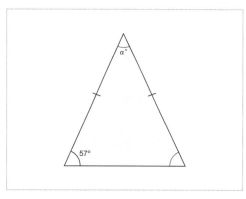

**12.** What is the size of angle *x*?
*Don't include the units in your answer.*

1
2
3 ▪ 106

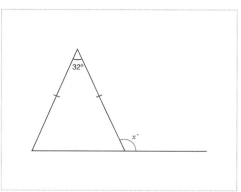

Level 2: *cont.*

**13.** Calculate the size of angle *d*.
*Don't include the units in your answer.*

1
2
3 ▪ 300

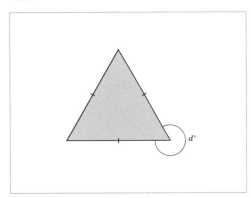

**14.** The value of angle *DEF* = _____.
Express your answer as an algebraic expression.

a
b
c ▪ 360 - (a + b + c) ▪ 360 - a - b - c

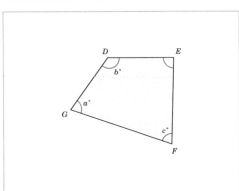

**15.** Calculate the value of *k*.

1
2
3 ▪ 45

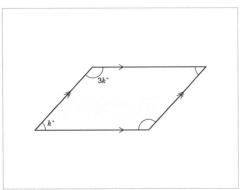

**16.** What is the value of *s*?

1
2
3 ▪ 24

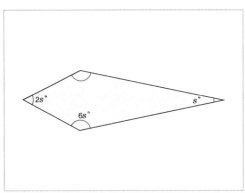

**17.** What is the value of *a*?

1
2
3 ▪ 15

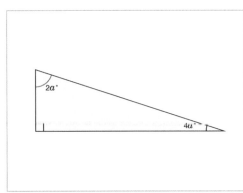

**18.** Calculate the size of angle *f*.
*Don't include the units in your answer.*

1
2
3 ▪ 114

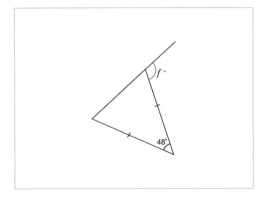

**Level 2:** *cont.*

**19.** The value of angle *XYZ* = _____.

**a b c** Express your answer as an algebraic expression.

- 180 - (a + b)  ▪ 180 - a - b

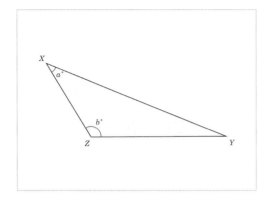

**20.** Calculate the value of *p*.

**1 2 3** ▪ 36

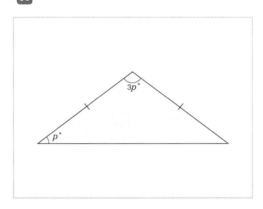

**Level 3:** Reasoning - Triangles with no measurements.

❋ **Required:** 5/5    ❋ **Student Navigation:** on
❋ **Randomised:** off

**21.** Explain why you can't draw a triangle with two
**a b c** obtuse angles.

**22.** Prove that the angles in a triangle add up to 180
**a b c** degrees.

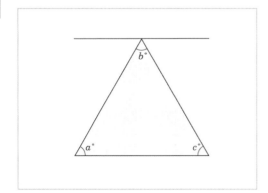

**23.** If one angle in the triangle is 72 degrees and the
**a b c** sum of the base angles is less than 110, what is
the size of the smallest angle?
*Don't include the units in your answer.*

- 54

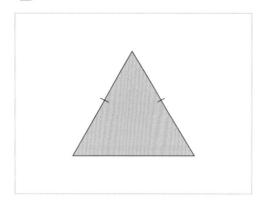

**24.** Jack says, "I can calculate angle *b* using the
**a b c** information I've been given".
Is Jack correct? Explain your answer.

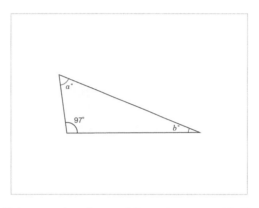

**25.** If three angles of a quadrilateral measure 40, 40
☐☒☐ and 50 degrees, what type of shape is the
quadrilateral?

1/7  ▪ **Rectangle** ▪ **Trapezium** ▪ Arrowhead ▪ **Kite**
▪ **Rhombus** ▪ **Square** ▪ **Parallelogram**

**Level 4:** Problem solving - Finding angles in composite
shapes.

❋ **Required:** 5/5    ❋ **Student Navigation:** on
❋ **Randomised:** off

**26.** Calculate angle *b*.
**1 2 3** *Don't include the units in your answer.*

- 75

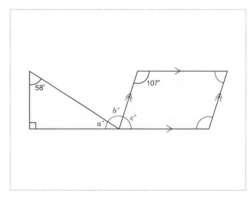

**27.** Calculate the size of angle *y*.

*Don't include the units in your answer.*

▪ 35

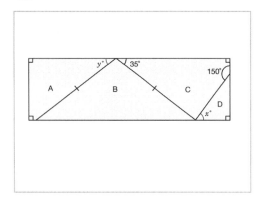

**28.** Calculate the size of angle *c*.

*Don't include the units in your answer.*

▪ 113

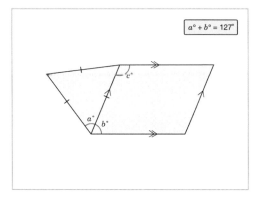

**29.** What is the the size of angle *y*?

*Don't include the units in your answer.*

▪ 108

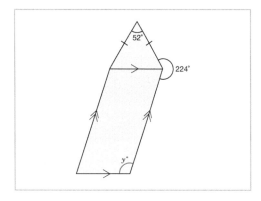

**30.** Calculate the size of angle *y*.

*Don't include the units in your answer.*

▪ 79

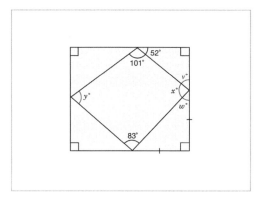

# Use alternate and corresponding angles on parallel lines

**Competency:**    Understand and use the relationship between parallel lines and alternate and corresponding angles.

**Quick Search Ref:**    10134

**Correct:** Correct.    **Wrong:** Incorrect, try again.    **Open:** Thank you.

**Level 1:**    Understanding - Identify alternate and corresponding angles.

✿ **Required:** 7/10    ✿ **Student Navigation:** on    ✿ **Randomised:** off

**1.**    Select the two parallel lines.

■ Line AB  ■ Line CD  ■ **Line EF**  ■ **Line GH**

2/4

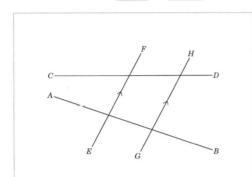

**2.**    A transversal is:

1/3
■ a line which is always the same distance apart from another line.
■ a straight line which crosses one other straight line.
■ **a straight line that crosses at least two other straight lines.**

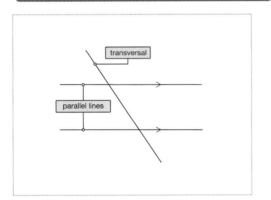

**3.**    Select the alternate angle to *a*.

■ Angle b  ■ **Angle c**  ■ Angle d

1/3

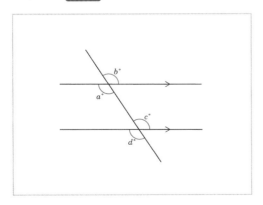

**4.**    Select the two pairs of alternate angles.

■ Angles a and e  ■ Angles b and c  ■ Angles b and d
■ **Angle b and f**  ■ **Angle c and e**  ■ Angles e and d

2/6

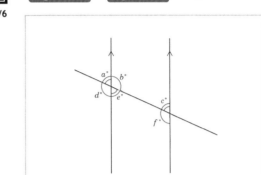

**5.**    Angles *b* and *f* are:

■ alternate angles.  ■ angles on a straight line.
■ **corresponding angles.**  ■ vertically opposite angles.

1/4

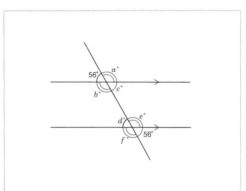

**6.**    Which is the corresponding angle to *a*?

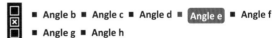
■ Angle b  ■ Angle c  ■ Angle d  ■ **Angle e**  ■ Angle f
■ Angle g  ■ Angle h

1/7

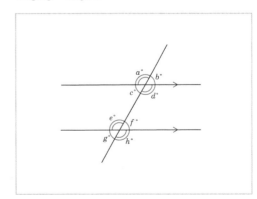

**Level 1:** *cont.*

**7.** Select the three angles which have the same value as angle *n*.

- Angle l ▪ Angle m ▪ Angle o ▪ Angle p ▪ Angle q
3/7 ▪ Angle r ▪ Angle s

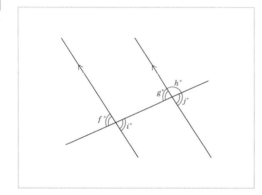

**8.** Select the alternate angle to *i*.

- Angle f ▪ Angle g ▪ Angle h ▪ Angle j
1/4

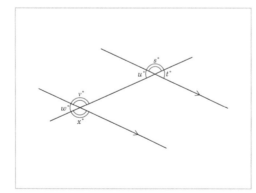

**9.** Which is the corresponding angle to *s*?

- Angle t ▪ Angle u ▪ Angle v ▪ Angle w ▪ Angle x
1/5

**10.** Select the three angles which have the same value as angle *g*.

- Angle a ▪ Angle b ▪ Angle c ▪ Angle d ▪ Angle e
3/7 ▪ Angle f ▪ Angle h

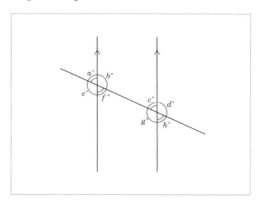

**Level 2:** Fluency - Use alternate and corresponding angles to calculate the value of missing angles.

✱ **Required:** 7/10   ✱ **Student Navigation:** on
✱ **Randomised:** off

**11.** What is the value of *x*?

▪ 93

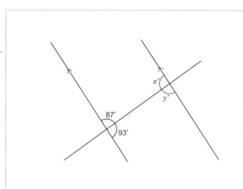

**12.** What is the value of *g*?

▪ 139

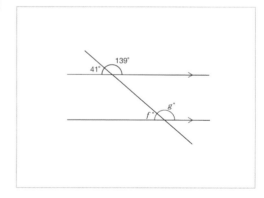

**Level 2: cont.**

**13.** Calculate the value of angle *x*.

 ▪ **77**

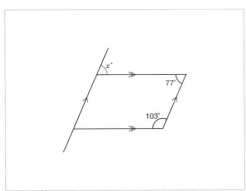

**14.** Calculate the value of *d*.

 ▪ **116**

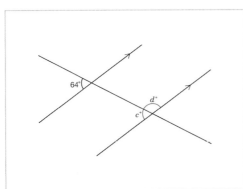

**15.** What is the value of *h*?

 ▪ **47**

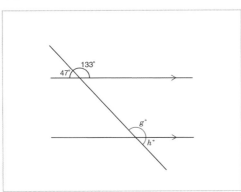

**16.** The value of *y* is ___°.

 ▪ **71**

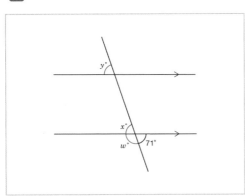

**17.** What is the value of *a*?

 ▪ **39**

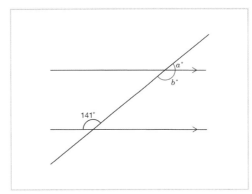

**18.** What is the value of *m*?

 ▪ **101**

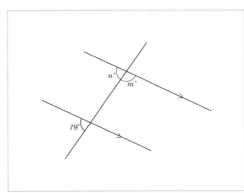

**19.** What is the value of *g*?

 ▪ **151**

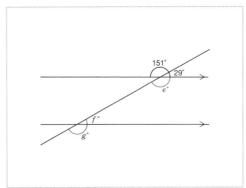

**20.** What is the value of *y*?

 ▪ **98**

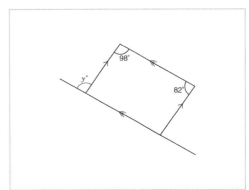

**Level 3:** Reasoning - Alternate and corresponding angles.

🟦 Required: 5/5   🟦 Student Navigation: off
🟦 Randomised: off

**21.** Kelly says, "Lines *AB* and *CD* are parallel because they run along side each other".
`a` `b` `c`   Is Kelly correct? Explain your answer.

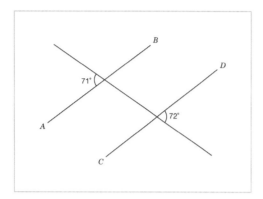

**22.** Calculate the value of *z*.

`1` `2` `3`  ▪ `108`

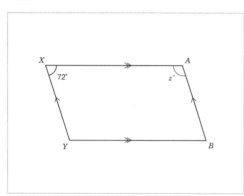

**23.** Vanessa says, "Angles *a* and *b* add up to 180 degrees". Prove Vanessa is correct by finding the sum of the angles in terms of *a*.
`a` `b` `c`

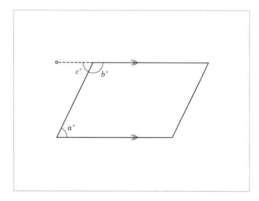

**24.** In the diagram, how many of the angles measure 47 degrees?

▪ 4 ▪ 5 ▪ 6 ▪ 7 ▪ `8`

1/5

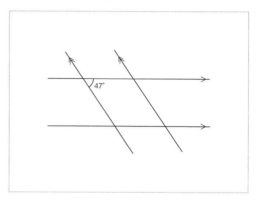

**25.** Calculate the value of *c*.

`1` `2` `3`  ▪ `127`

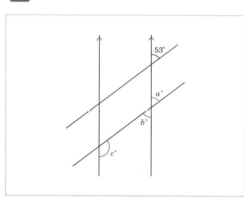

**Level 4:** Problem Solving - Calculating the size of angles.

🟦 Required: 5/5   🟦 Student Navigation: on
🟦 Randomised: off

**26.** Calculate the size of angle *BCD*.
`1` `2` `3`   *Don't include the units in your answer.*

▪ `87`

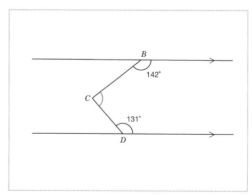

**27.** What is the size of angle *XYZ*?

*Don't include the units in your answer.*

1
2
3  ▪ 42

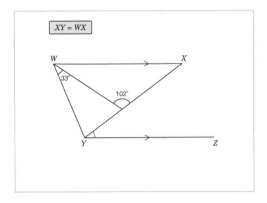

**28.** Calculate the size of angle *x*.

*Don't include the units in your answer.*

1
2
3  ▪ 96

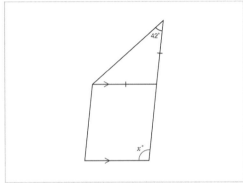

**29.** What is the size of angle *a*?

*Don't include the units in your answer.*

1
2
3  ▪ 14

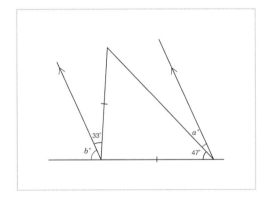

**30.** Calculate the value of *x*.

*Don't include the units in your answer.*

1
2
3  ▪ 43

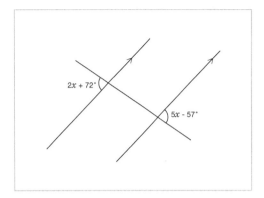

# Use the sum of angles in a triangle to calculate the angle sum in any polygon, and to derive properties of regular polygons

**Competency:** Derive and use the sum of angles in a triangle and use it to deduce the angle sum in any polygon, and to derive properties of regular polygons and a quadrilateral.

**Quick Search Ref:** 10114

Correct: Correct. Wrong: Incorrect, try again. Open: Thank you.

**Level 1:** Understanding - Using triangles to find angles in a polygon.

✿ Required: 7/10 ✿ Student Navigation: on ✿ Randomised: off

**1.** Select the three statements which are true.

- All angles in a regular polygon are equal.
- All angles in an irregular polygon are equal.

3/5
- An irregular polygon can have angles which are all different sizes.
- The sum of all angles in a triangle is 360 degrees.
- The sum of all angles in a triangle is 180 degrees.

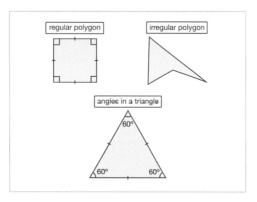

**2.** Which method shows the simplest way to calculate the sum of all angles in an irregular quadrilateral?

1/2 ■ (i) ■ (ii)

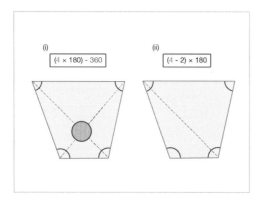

**3.** The hexagon has been divided into the **least possible number of triangles**. Which calculation would you use to calculate the sum of the interior angles in the hexagon?

- 4 × 180 = 720 ■ 6 × 180 = 1,080 ■ 4 × 360 = 1,440
- 6 × 360 = 2,160

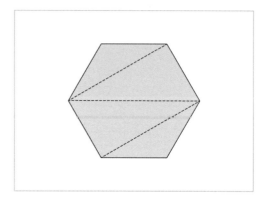

**4.** What is the least number of triangles the pentagon can be divided into?

1
2
3 ■ 3

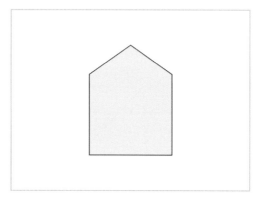

**5.** Calculate the sum of the interior angles in the pentagon.
*Don't include the units in your answer.*

- 540

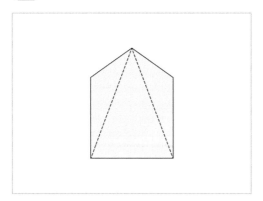

**6.** What Is the size of angle *a* in the regular pentagon?
*Don't include the units in your answer.*

- 108

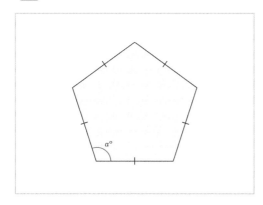

**7.** What is the size of one exterior angle in a regular octagon?
*Don't include the units in your answer.*

- 45

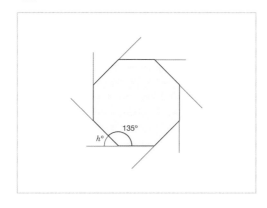

**8.** What is the least number of triangles the octagon can be split into?
*Don't include the units in your answer.*

- 6

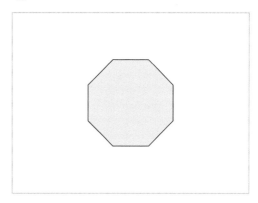

**9.** Calculate the sum of the interior angles in the heptagon.
*Don't include the units in your answer.*

- 900

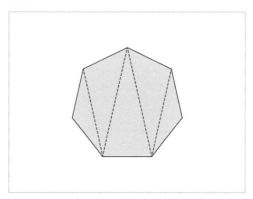

**10.** What is the size of angle *x* in the regular hexagon?
*Don't include the units in your answer.*

- 120

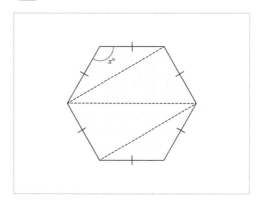

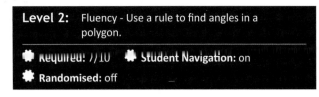

**Level 2:** Fluency - Use a rule to find angles in a polygon.

Required: 7/10   Student Navigation: on
Randomised: off

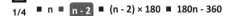

**11.** Which rule can you use to find the least number of triangles in a polygon, where *n* is the number of sides?

1/4   ▪ n ▪ n - 2 ▪ (n - 2) × 180 ▪ 180n - 360

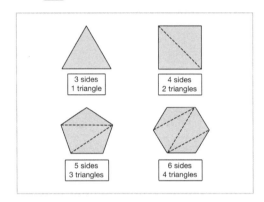

3 sides
1 triangle

4 sides
2 triangles

5 sides
3 triangles

6 sides
4 triangles

**12.** What is the least number of triangles in a dodecagon?

▪ 10

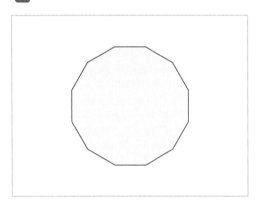

**13.** Select the two formulae you can use to calculate the sum of the interior angles in a polygon, where *n* is the number of sides.

2/6   ▪ 180n ▪ 180n - 360 ▪ (n - 2) - 360 ▪ (n - 2) × 180
▪ (n - 2) - 180 ▪ (n - 2) × 360

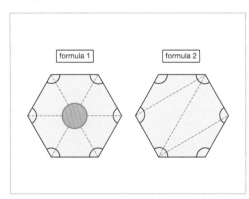

formula 1          formula 2

**14.** Calculate the sum of the angles in the polygon.
*Don't include the units in your answer.*

a
b
c

▪ 1260 ▪ 1,260

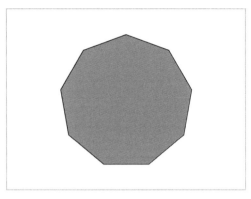

**15.** What is the size of one interior angle in the regular decagon?
*Don't include the units in your answer.*

a
b
c

▪ 144

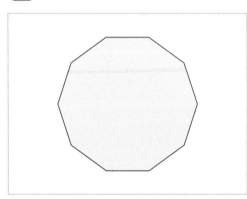

**16.** Calculate the size of angle *x*.
*Don't include the units in your answer.*

1
2
3

▪ 245

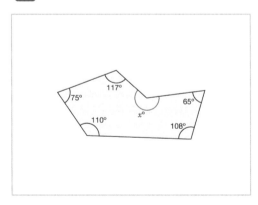

117°
75°
*x*°
65°
110°
108°

**17.** Sort angles *a* to *d* in ascending order of size (smallest angle first).

↑
↓ ▪ Angle c ▪ Angle a ▪ Angle d ▪ Angle b

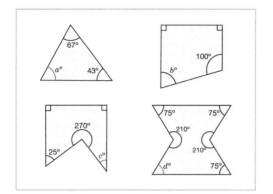

**18.** Calculate the sum of the angles in the polygon.

a
b
c

*Don't include the units in your answer.*

▪ 900

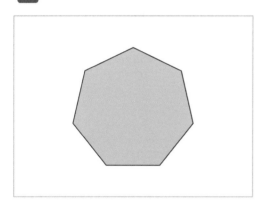

**19.** What is the size of angle *f*?

1
2
3

*Don't include the units in your answer.*

▪ 42

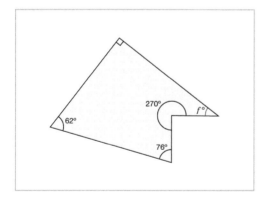

**20.** Sort angles *w* to *z* in ascending order of size (smallest angle first).

↑
↓ ▪ Angle w ▪ Angle y ▪ Angle x ▪ Angle z

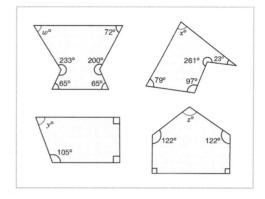

**Level 3:** Reasoning - Calculations involving interior and exterior angles of polygons.

✱ **Required:** 5/5  ✱ **Student Navigation:** on
✱ **Randomised:** off

**21.** Kai says, "The sum of all angles in a regular hexagon is 1,080 degrees".

a
b
c

Explain Kai's mistake.

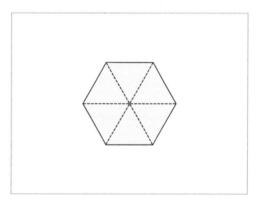

**22.** Calculate the size of one exterior angle in a regular nonagon.

a
b
c

▪ 40

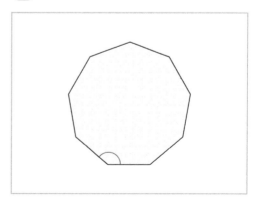

**23.** Debbie says, "Each angle in the pentagon equals
a
b
c    108 degrees".
Is Debbie correct? Explain your answer.

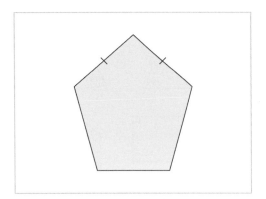

**24.** Polygons tesselate when they connect around a
a
b
c    point without leaving any gaps. Explain why three
regular pentagons do not tesselate.

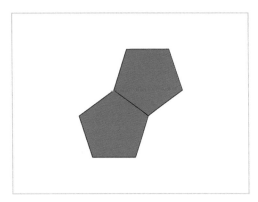

**25.** The sum of all exterior angles in a polygon equals
☐
☒
☐    360 degrees. Which two formulae can be used to
calculate the size of one interior angle in a regular
polygon?

2/4

■ 180 - ((180n - 360) ÷ n)    ■ 180 - (360 ÷ n)    ■ 360 ÷ n

■ (180n - 360) ÷ n

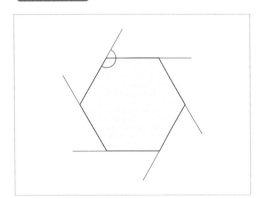

✱ **Required:** 5/5    ✱ **Student Navigation:** on
✱ **Randomised:** off

**26.** What is the value of angle *x*?
1
2
3    *Don't include the units in your answer.*

■ 85

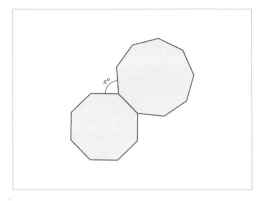

**27.** Calculate the size of angle *DEF*.
1
2
3    *Don't include the units in your answer.*

■ 115

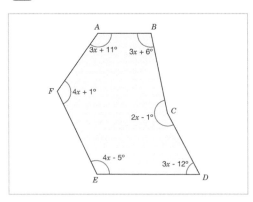

**28.** If one interior angle of a regular polygon is 165
a
b
c    degrees, what is the sum of the polygon's interior
angles?

■ 3960    ■ 3,960

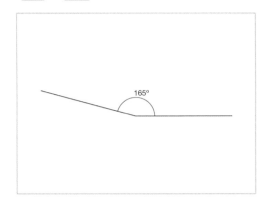

**29.** Jenny tiles her kitchen floor using two different
a
b
c    shaped tiles. Both are regular polygons and cover
the floor without leaving any gaps. If one of the
shapes is octagonal, what shape is the other tile?

■ a square    ■ squares    ■ Square

**30.** One exterior angle of a regular polygon is 5

degrees. What is the sum of the polygon's interior
angles?

- 12600  - 12,600

# Identify reflective and rotational symmetry of polygons

**Competency:** Describe, sketch and draw using conventional terms and notations: points, lines, parallel lines, perpendicular lines, right angles, regular polygons, and other polygons that are reflectively and rotationally symmetrical.

**Quick Search Ref:** 10303

Correct: Correct.    Wrong: Incorrect, try again.    Open: Thank you.

**Level 1:** Understanding - Identifying reflective and rotational symmetry.

✼ Required: 7/10    ✼ Student Navigation: on    ✼ Randomised: off

**1.** Select the statement that describes rotational symmetry.

1/3

- ■ The number of positions a shape can be rotated to between 0° to 360° around its centre and still look the same.
- ■ When an imaginary line divides a shape into two mirror image halves.
- ■ When a shape still looks the same after a rotation between 0° and 360°.

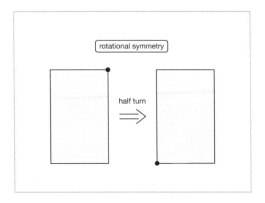

**2.** Which statement describes reflective symmetry?

1/3

- ■ When a shape still looks the same after a rotation between 0° and 360° around its centre.
- ■ The number of positions a shape can be rotated to between 0° and 360° around its centre and still look the same.
- ■ When an imaginary line divides a shape into two mirror image halves.

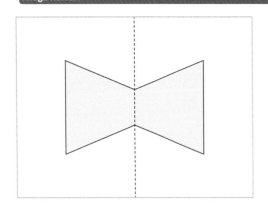

**3.** Does the image show reflective or rotational symmetry?

1/2

- ■ rotational symmetry ■ reflective symmetry

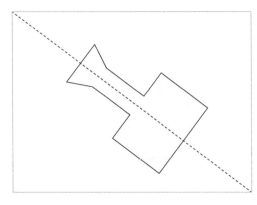

**4.** Which two images show reflective symmetry?

2/4

- ■ (i) ■ (ii) ■ (iii) ■ (iv)

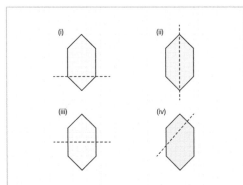

**5.** Which shape has only two lines of reflective symmetry?

1/4

- ■ (i) ■ (ii) ■ (iii) ■ (iv)

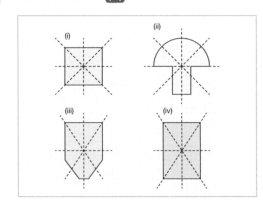

**6.** Which two examples show rotational symmetry of shape A?

2/4

■ (i) ■ (ii) ■ (iii) ■ (iv)

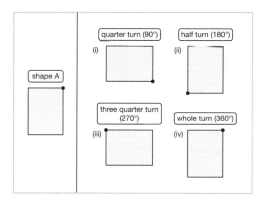

**7.** An equilateral triangle has rotational symmetry of order __.

1/5

■ 2 ■ 3 ■ 4 ■ 5 ■ 6

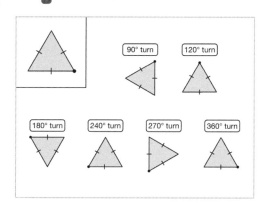

**8.** A square has rotational symmetry of order __.

1
2
3

■ 4

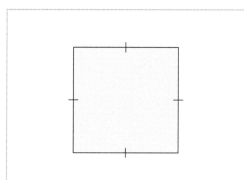

**9.** How many lines of reflective symmetry does this shape have?

1
2
3

■ 2

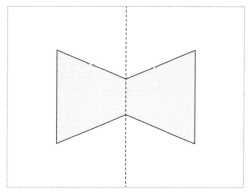

**10.** Which shape has only two lines of reflective symmetry?

1/4

■ (i) ■ (ii) ■ (iii) ■ (iv)

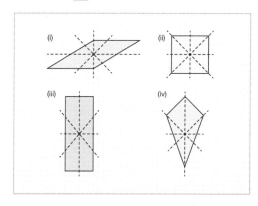

**Level 2:** Fluency - Identifying total lines of symmetry and rotational order.

✿ **Required:** 7/10  ✿ **Student Navigation:** on
✿ **Randomised:** off

**11.** How many lines of reflective symmetry does the shape have?

a
b
c

■ 1

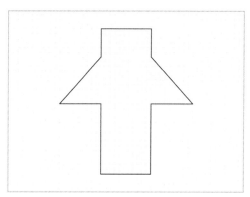

**Level 2:** *cont.*

**12.** Which two polygons have five or more lines of reflective symmetry?

■ (i) ■ (ii) ■ (iii) ■ (iv)

2/4

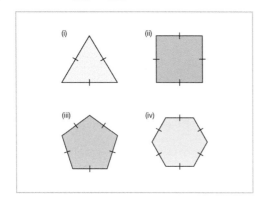

**13.** What is the order of rotational symmetry of a regular pentagon?

■ 5

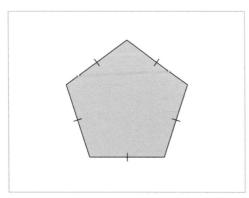

**14.** Select the polygon that has the same order of rotational symmetry as a rhombus.

■ (i) ■ (ii) ■ (iii) ■ (iv)

1/4

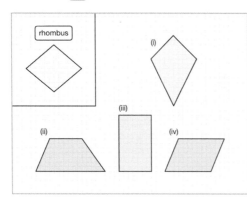

**15.** How many lines of symmetry does a parallelogram have?

■ 0

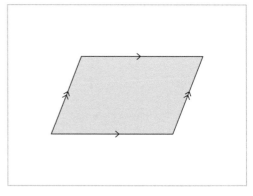

**16.** What is the order of rotational symmetry of the dodecagon?

■ 4

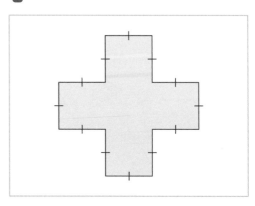

**17.** How many lines of symmetry does a circle have?

■ 2 ■ 4 ■ 360 ■ infinite

1/4

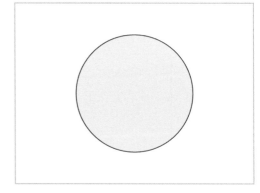

**18.** What is the order of rotational symmetry of the octagon?

 ▪ 2

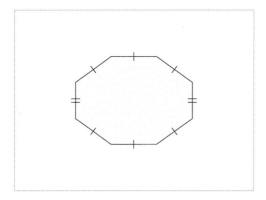

**19.** How many lines of symmetry does a rhombus have?

 ▪ 2

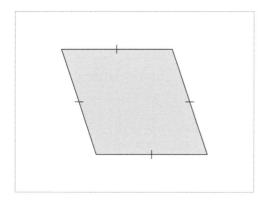

**20.** Select the polygon which has the same order of rotational symmetry as a rectangle.

 ▪ (i) ▪ (ii) ▪ (iii) ▪ (iv)

1/4

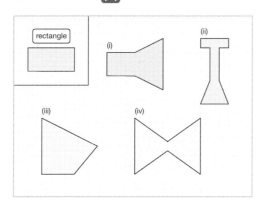

**Level 3:** Reasoning - Identifying total number lines of symmetry and rotational order in the same shape.

✿ **Required:** 5/5     ✿ **Student Navigation:** on
✿ **Randomised:** off

**21.** What is the smallest number of degrees you need to rotate a regular pentagon by to get rotational symmetry?
*Don't include the units in your answer.*

▪ 72

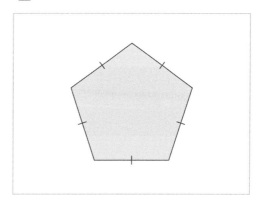

**22.** Select the shapes which go in box D.

 ▪ (i) ▪ (ii) ▪ (iii) ▪ (iv)

2/4

| | no rotational symmetry | rotational symmetry |
|---|---|---|
| no reflective symmetry | A | B |
| reflective symmetry | C | D |

(i)          (ii)          (iii)          (iv)

**23.** Tony says, "A square has rotational symmetry of order 4 and four lines of symmetry. So, all quadrilaterals must have the same order of rotational symmetry as lines of symmetry".
Is Tony correct? Explain your answer.

**24.** How many lines of symmetry are there in a regular *n*-sided shape?

 ▪ n

**Level 3: cont.**

**25.** Nathan says, "The octagon has rotational
a
b  symmetry of order 8".
c  Is Nathan correct? Explain your answer.

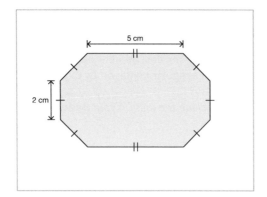

**Level 4:**   Problem Solving - Questions in context.

✱ **Required:** 5/5   ✱ **Student Navigation:** on
✱ **Randomised:** off

**26.** How many different ways can the jigsaw piece be
1
2  placed in the board?
3
   ▪ **4**

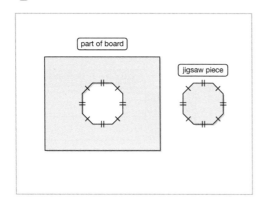

**27.** Gina has 50 square pieces of paper. What is the
1
2  greatest number of squares she can use to make a
3  shape with four lines of reflective symmetry?

   ▪ **49**

**28.** If two cogs start to rotate and then stop the next
1
2  time A and B touch, how many full rotations will
3  cog A make before the cogs come to a stop?
   *Both cogs are regular polygons.*

   ▪ **3**

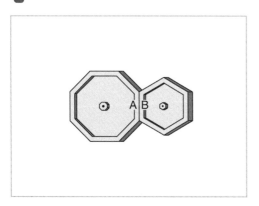

**29.** What is the minimum number of squares Jenny
1
2  needs to shade to give the shape rotational
3  symmetry of order 2?

   ▪ **5**

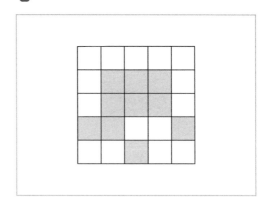

**30.** How many ways can you arrange the tiles in the
1
2  grid to make a shape with reflective symmetry?
3
   ▪ **14**

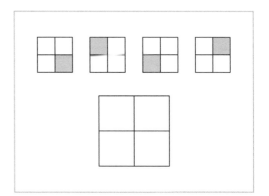

# Mathematics

## Y7

## Statistics

Tables

Bar Charts

Pictograms

Line Graphs

Pie Charts

Averages and Range

# Complete, read and interpret frequency tables and two-way tables

**Competency:** Construct and interpret appropriate tables, charts and diagrams, including frequency tables, bar charts, pie charts and pictograms for categorical data and vertical line (or bar) charts for ungrouped and grouped data.

**Quick Search Ref:** 10225

Correct: Correct.    Wrong: Incorrect, try again.    Open: Thank you.

**Level 1:** Understanding - Complete and read data from a frequency table and a two-way table.

✳ **Required:** 7/10    ✳ **Student Navigation:** on    ✳ **Randomised:** off

**1.** In the following set of ungrouped data, what is the frequency of the number 4?

1, 3, 6, 4, 5, 6, 3, 4, 6, 3, 6.

▪ **2**

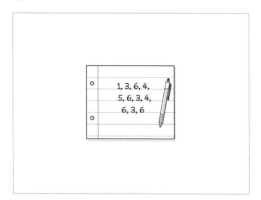

**2.** Which number fills both of the blanks in the frequency table?

▪ 3 ▪ **4** ▪ 5 ▪ 6

1/4

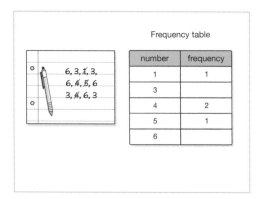

Frequency table

| number | frequency |
|--------|-----------|
| 1 | 1 |
| 3 | |
| 4 | 2 |
| 5 | 1 |
| 6 | |

**3.** Use the ungrouped data to find the missing number in the frequency table.

▪ **6**

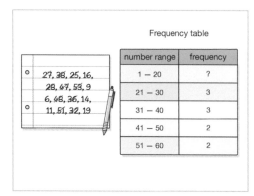

Frequency table

| number range | frequency |
|--------------|-----------|
| 1 — 20 | ? |
| 21 — 30 | 3 |
| 31 — 40 | 3 |
| 41 — 50 | 2 |
| 51 — 60 | 2 |

**4.** Nine children go to Harry's birthday party. Enter the remaining ungrouped data into the table and select the most common age range.

1/5  ▪ 1 - 2 years old ▪ 3 - 4 years old ▪ 5 - 6 years old
▪ 7 - 8 years old ▪ 9 - 10 years old

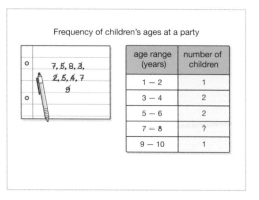

Frequency of children's ages at a party

| age range (years) | number of children |
|-------------------|--------------------|
| 1 — 2 | 1 |
| 3 — 4 | 2 |
| 5 — 6 | 2 |
| 7 — 8 | ? |
| 9 — 10 | 1 |

**5.** The total snowfall in a number of different cities is recorded in a frequency table. How many cities have between 13 and 15 inches of snow?

▪ **4**

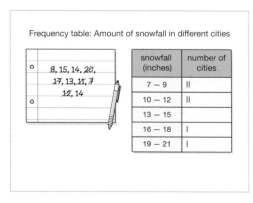

Frequency table: Amount of snowfall in different cities

| snowfall (inches) | number of cities |
|-------------------|------------------|
| 7 — 9 | II |
| 10 — 12 | II |
| 13 — 15 | |
| 16 — 18 | I |
| 19 — 21 | I |

**6.** According the table, how many people in the office drink tea with sugar?

▪ **6**

Frequency of hot drink preferences

| | no sugar | with sugar | total: |
|---------|----------|------------|--------|
| tea | ? | ? | ? |
| coffee | 9 | 3 | 12 |
| total: | 18 | 9 | 27 |

**7.** How many season ticket holders are male?

 ▪ **31**

Frequency of season ticket holders by age and gender

|  | male | female | total: |
|---|---|---|---|
| under 16 |  | 25 | 42 |
| 16 or over |  | 32 | 46 |
| total: | ? |  |  |

**8.** The number of goals scored by Will's football team is recorded. What range of goals does Will's team score the most often?

1/4    ▪ **0 - 2 goals**   ▪ 3 - 4 goals   ▪ 5 - 6 goals   ▪ 7 - 8 goals

Frequency distribution of goal range per match

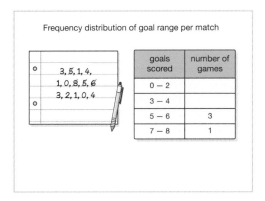

3, 5, 1, 4,
1, 0, 8, 5, 6
3, 2, 1, 0, 4

| goals scored | number of games |
|---|---|
| 0 − 2 |  |
| 3 − 4 |  |
| 5 − 6 | 3 |
| 7 − 8 | 1 |

**9.** How many children in the class have brown eyes?

 ▪ **11**

Distribution of eye colour by gender

|  | boys | girls | total: |
|---|---|---|---|
| green | 7 | 8 | 15 |
| blue | 5 | 4 | 9 |
| brown | 4 | 7 | ? |
| total: | 16 | 19 | 35 |

**10.** A survey asked people whether they prefer white chocolate, milk chocolate or dark chocolate. How many females took part in the survey?

▪ **27**

Frequency distribution of chocolate preferences by gender

|  | white | milk | dark | total: |
|---|---|---|---|---|
| females | 7 | 12 | 8 | ? |
| males | 10 | 21 | 5 | 36 |
| total: | 17 | 33 | 13 | 63 |

**Level 2:**   Fluency - Finding the sum, difference or total in a frequency table and two-way table.

✿ **Required:** 7/10   ✿ **Student Navigation:** on
✿ **Randomised:** off

**11.** A gym logs the heart rate data of its members during their workouts. How many more members record a heart rate of 111 - 130 beats per minute (b.p.m.) than 71 - 90 b.p.m.?

▪ **29**

Frequency distribution of heart rate ranges

| heart rate (b.p.m.) | number of members |
|---|---|
| 61 − 70 | 7 |
| 71 − 80 | 27 |
| 81 − 90 | 34 |
| 91 − 100 | 19 |
| 101 − 110 | 27 |
| 111 − 120 | 32 |
| 121 − 130 | 58 |
| 131 − 140 | 9 |

**12.** How many more Year 7 pupils than Year 11 pupils have school dinners?

 ▪ **23**

Distribution of school dinner preferences

|  | year 7 | year 8 | year 9 | year 10 | year 11 | total: |
|---|---|---|---|---|---|---|
| school dinners | ? | ? | ? | ? | ? | 381 |
| packed lunches | 35 | 52 | 45 | 36 | 72 | 240 |
| total: | 121 | 137 | 117 | 111 | 135 | 621 |

**13.** How many more people watch television for 0 - 4 hours a day than for 5 - 8 hours a day?

1
2
3   ▪ 32

Distribution of TV viewing time ranges

| time (hours) | number of people |
|---|---|
| 0 — 2 | 27 |
| 3 — 4 | 44 |
| 5 — 6 | 21 |
| 7 — 8 | 18 |

**14.** Each pupil in class 7b is asked how many coloured pens they have in their pencil case. What range of coloured pens is the most common?

1/6   ▪ 0 - 5   ▪ 6 - 10   ▪ 11 - 15   ▪ 16 - 20   ▪ 21 - 25   ▪ 26 - 30

Distribution of coloured-pen ownership

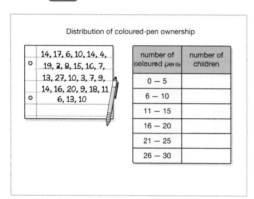

14, 17, 6, 10, 14, 4, 19, 2, 8, 15, 16, 7, 13, 27, 10, 3, 7, 9, 14, 16, 20, 9, 18, 11, 6, 13, 10

| number of coloured pens | number of children |
|---|---|
| 0 — 5 | |
| 6 — 10 | |
| 11 — 15 | |
| 16 — 20 | |
| 21 — 25 | |
| 26 — 30 | |

**15.** 91 Year 7 pupils were asked to choose their favourite sport. Arrange the categories in descending order (highest first).

▪ girls who play football   ▪ boys who play rugby
▪ boys who play hockey   ▪ girls who play rugby

Distribution of sport preferences by gender

| | football | rugby | hockey | total: |
|---|---|---|---|---|
| boys | 33 | | | 52 |
| girls | | | 19 | |
| total: | | 13 | 28 | 91 |

**16.** A shop records how much money each customer spends to the nearest pound (£). What is the most common spending range?

1/6   ▪ 0 - 5   ▪ 6 - 10   ▪ 11 - 15   ▪ 16 - 20   ▪ 21 - 25   ▪ 26 - 30

Frequency of spending range per customer

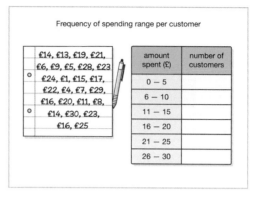

£14, £13, £19, £21, £6, £9, £5, £28, £23, £24, £1, £15, £17, £22, £4, £7, £29, £16, £20, £11, £8, £14, £30, £23, £16, £25

| amount spent (£) | number of customers |
|---|---|
| 0 — 5 | |
| 6 — 10 | |
| 11 — 15 | |
| 16 — 20 | |
| 21 — 25 | |
| 26 — 30 | |

**17.** A geography class is asked whether they have been to Spain, Greece or France. How many more children have been to Spain than Greece?

1
2
3   ▪ 33

Frequency distribution of countries visited by gender

| | Spain | Greece | France | total: |
|---|---|---|---|---|
| boys | | 27 | | 80 |
| girls | | | 7 | 58 |
| total: | | 42 | 21 | |

**18.** How many flights cost £100 or more?

1
2
3   ▪ 111

Distribution of flights sold by price range

| cost of flight (£) | number of flights |
|---|---|
| 0 — 49 | 13 |
| 50 — 99 | 18 |
| 100 — 149 | 27 |
| 150 — 199 | 32 |
| 200 — 250 | 52 |

**19.** How many more women entered the swimming competition than men?

- ■ 13

Distribution of swimming competition entrants by gender

|  | front crawl | breast stroke | back stroke | butterfly | total: |
|---|---|---|---|---|---|
| male |  | 10 |  | 12 |  |
| female | 21 |  | 17 |  |  |
| total: | 39 |  | 36 | 31 | 131 |

**20.** 72 children are asked what pets they own. Arrange the categories in ascending order (smallest first).

- ■ boys with a rabbit ■ boys with a dog
- ■ girls with a dog ■ girls with a cat

Frequency distribution of pets owned by gender

|  | cat | dog | rabbit | total: |
|---|---|---|---|---|
| boys | 7 |  |  | 34 |
| girls |  |  | 5 |  |
| total: |  | 31 | 17 | 72 |

**Level 3:** Reasoning - Reading frequency tables and two-way tables which require amendments.

✿ **Required:** 5/5    ✿ **Student Navigation:** on
✿ **Randomised:** off

**21.** Billy has completed the two-way table using the data shown. Explain what mistake he has made.

Frequency distribution of shapes by colour

|  | ○ | □ | △ | total: |
|---|---|---|---|---|
| red | 2 | 4 | 1 | 7 |
| blue | 3 | 2 | 3 | 8 |
| green | 1 | 1 | 3 | 5 |
| total: | 6 | 7 | 7 |  |

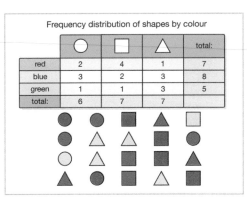

**22.** Temporary traffic lights mean every bus journey is taking 18 minutes longer. What is the first bus that will arrive at Finchley after 10:00?

1/5    ■ bus A ■ bus B ■ bus C ■ bus D ■ bus E

Bus times from Keenal to Brent

|  | bus A | bus B | bus C | bus D | bus E |
|---|---|---|---|---|---|
| Keenal | 07:42 | 07:59 | 08:37 | 09:06 | 09:31 |
| Tintou | 08:14 | 08:31 | 09:04 | 09:35 | 10:03 |
| Finchley | 08:31 | 08:50 | 09:25 | 09:52 | 10:21 |
| Brent | 08:47 | 09:05 | 09:39 | 10:01 | 10:33 |

**23.** Sam lives in Soweray and works in Hepton. It takes her 12 minutes to walk from home to Soweray train station and another 15 minutes to walk from Hepton train station to work. What is the latest time Sam can leave her house if she starts work at 11:30?

- ■ 10:24

Train times from Soweray to Hepton

| Soweray | 10:15 | 10:31 | 10:36 | 10:40 | 10:48 |
|---|---|---|---|---|---|
| Trafax | 10:33 | 10:45 | 10:51 | 10:56 | 11:02 |
| Foulder | 10:46 | 10:53 | 10:59 | 11:05 | 11:13 |
| Hepton | 10:59 | 11:05 | 11:12 | 11:18 | 11:24 |

**24.** How many goals were scored in the league over the weekend?

- ■ 129

Frequency of goals scored in the league

| goals scored | number of games | total: |
|---|---|---|
| 0 | 6 |  |
| 1 | 4 |  |
| 2 | 9 |  |
| 3 | 11 |  |
| 4 | 13 |  |
| 5 | 2 |  |
| 6 | 1 |  |
| 7 | 0 |  |

**25.** According to the table, how many children in Year 7 have two siblings?

1 2 3  ▪ 91

Frequency distribution of siblings

|  |  | number of sisters | | |
|---|---|---|---|---|
|  |  | 0 | 1 | 2 |
| brothers | 0 | 13 | 19 | 28 |
| | 1 | 18 | 26 | 22 |
| | 2 | 37 | 21 | 17 |

**Level 4:**  Problem Solving - Complete frequency tables and two-way tables.

🟊 **Required:** 5/5   🟊 **Student Navigation:** on

🟊 **Randomised:** off

**26.** 35 children are asked whether they would prefer to go to a science museum or theme park. How many boys want to go to the science museum?
11 children said science museum.
12 girls said theme park.
There are 17 boys in total.

1 2 3

▪ 5

Distribution of school trip preferences by gender

|  | boys | girls | total: |
|---|---|---|---|
| science museum |  |  |  |
| theme park |  |  |  |
| total: |  |  |  |

**27.** All children who attend the cinema are accompanied by an adult. What is the mean number of children per adult?

1 2 3

▪ 2.4

Frequency distribution of children per adult

| number of children per adult | number of adults | total number of children |
|---|---|---|
| 1 | 5 |  |
| 2 | 7 |  |
| 3 | 4 |  |
| 4 | 3 |  |
| 5 | 1 |  |
| total: |  |  |

**28.** What percentage of darts players in the tournament are left handed?
*Don't include the % sign in your answer.*

1 2 3

▪ 25

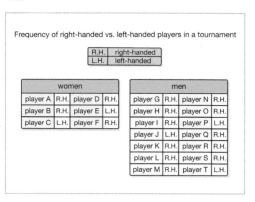

Frequency of right-handed vs. left-handed players in a tournament

| R.H. | right-handed |
| L.H. | left-handed |

| women | | | |
|---|---|---|---|
| player A | R.H. | player D | R.H. |
| player B | R.H. | player E | L.H. |
| player C | L.H. | player F | R.H. |

| men | | | |
|---|---|---|---|
| player G | R.H. | player N | R.H. |
| player H | R.H. | player O | R.H. |
| player I | R.H. | player P | L.H. |
| player J | L.H. | player Q | R.H. |
| player K | R.H. | player R | R.H. |
| player L | R.H. | player S | R.H. |
| player M | R.H. | player T | L.H. |

**29.** 418 children are asked which sport they would prefer. If 186 of the children asked were girls and 172 children said rugby, how many girls preferred rugby?
44 boys said hockey.
65 boys said tennis.
78 girls said tennis.

1 2 3

▪ 49

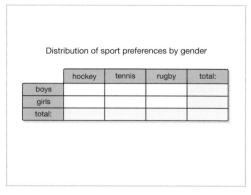

Distribution of sport preferences by gender

|  | hockey | tennis | rugby | total: |
|---|---|---|---|---|
| boys |  |  |  |  |
| girls |  |  |  |  |
| total: |  |  |  |  |

**30.** The owner of a cake stall keeps a list of how many cakes each customer buys. What is the mean number of cakes it sells to each customer?

1 2 3

▪ 3.25

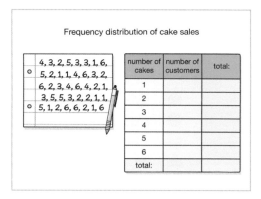

Frequency distribution of cake sales

| number of cakes | number of customers | total: |
|---|---|---|
| 1 |  |  |
| 2 |  |  |
| 3 |  |  |
| 4 |  |  |
| 5 |  |  |
| 6 |  |  |
| total: |  |  |

4, 3, 2, 5, 3, 3, 1, 6, 5, 2, 1, 1, 4, 6, 3, 2, 6, 2, 3, 4, 6, 4, 2, 1, 3, 5, 5, 3, 2, 2, 1, 1, 5, 1, 2, 6, 6, 2, 1, 6

# Complete, Read and Interpret Frequency Tables

| Competency: | Construct and interpret frequency tables for ungrouped and grouped data. |
|---|---|
| Quick Search Ref: | 10321 |

**Correct:** Correct.    **Wrong:** Incorrect, try again.    **Open:** Thank you.

**Level 1:**   Understanding - Complete and read data from a frequency table.

✿ **Required:** 7/10    ✿ **Student Navigation:** on    ✿ **Randomised:** off

**1.** In the following set of ungrouped data, what is the frequency of the number 4?
1, 3, 6, 4, 5, 6, 3, 4, 6, 3, 6.

▪ 2

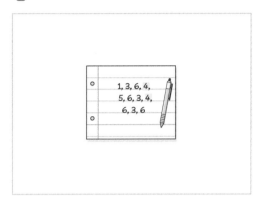

**2.** Which number fills both of the blanks in the frequency table?

▪ 3 ▪ 4 ▪ 5 ▪ 6

1/4

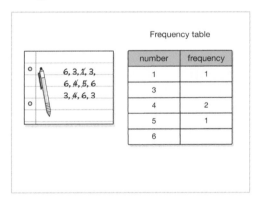

**3.** Use the ungrouped data to find the missing number in the frequency table.

▪ 6

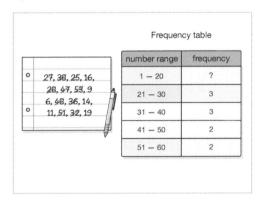

**4.** Nine children go to Harry's birthday party. Enter the remaining ungrouped data into the table and select the most common age range.

1/5

▪ 1 - 2 years old ▪ 3 - 4 years old ▪ 5 - 6 years old
▪ 7 - 8 years old ▪ 9 - 10 years old

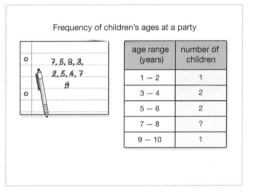

**5.** The total snowfall in a number of different cities is recorded in a frequency table. How many cities have between 13 and 15 inches of snow?

▪ 4

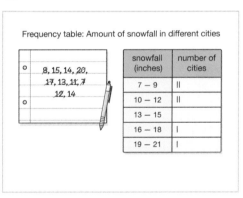

**6.** How many people drink tea in the morning?

▪ 6

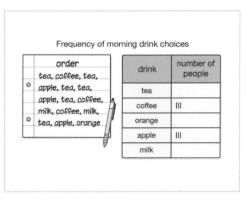

**7.** Class 6 took a test and the results are recorded. How many children scored 8 or more on the test?

[1 2 3]  ▪ 12

Frequency distribution of scores on a test

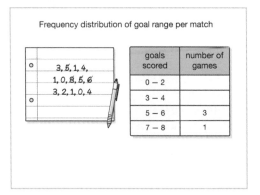

1, 8, 8, 2, 4,
5, 6, 9, 9, 10
3, 1, 4, 8, 10
3, 9, 10, 8, 8,
7, 8

| score | number of children |
|-------|---------------------|
| 1 | 2 |
| 2 | 1 |
| 3 |  |
| 4 |  |
| 5 |  |
| 6 |  |
| 7 |  |
| 8 |  |
| 9 |  |
| 10 | 3 |

**8.** The number of goals scored by Will's football team is recorded. What range of goals does Will's team score the most often?

1/4  ▪ **0 - 2 goals**  ▪ **3 - 4 goals**  ▪ **5 - 6 goals**  ▪ **7 - 8 goals**

Frequency distribution of goal range per match

3, 5, 1, 4,
1, 0, 8, 5, 6
3, 2, 1, 0, 4

| goals scored | number of games |
|--------------|-----------------|
| 0 – 2 |  |
| 3 – 4 |  |
| 5 – 6 | 3 |
| 7 – 8 | 1 |

**9.** Mr Hopkins records the types of birds he sees in his garden. How many sparrows did he see?

[1 2 3]  ▪ 11

Frequency distribution of bird sightings

| type of bird | tally | frequency |
|--------------|-------|-----------|
| bluebird | IIII IIII |  |
| hawk | I |  |
| robin | IIII IIII IIII I |  |
| sparrow | IIII IIII I |  |
| owl | II |  |

**10.** The total snowfall in a number of different cities is recorded in a frequency table. How many cities have between 10 and 15 inches of snow?

[1 2 3]  ▪ 6

Frequency table: Amount of snowfall in different cities

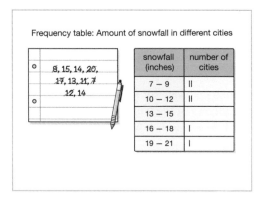

8, 15, 14, 20,
17, 13, 11, 7
12, 14

| snowfall (inches) | number of cities |
|-------------------|------------------|
| 7 – 9 | II |
| 10 – 12 | II |
| 13 – 15 |  |
| 16 – 18 | I |
| 19 – 21 | I |

**Level 2:** Fluency - Finding the sum, difference or total in a frequency table.

❊ **Required:** 7/10    ❊ **Student Navigation:** on
❊ **Randomised:** off

**11.** A gym logs the heart rate data of its members during their workouts. How many more members record a heart rate of 111 - 130 beats per minute (b.p.m.) than 71 - 90 b.p.m.?

[1 2 3]  ▪ 29

Frequency distribution of heart rate ranges

| heart rate (b.p.m.) | number of members |
|---------------------|-------------------|
| 61 – 70 | 7 |
| 71 – 80 | 27 |
| 81 – 90 | 34 |
| 91 – 100 | 19 |
| 101 – 110 | 27 |
| 111 – 120 | 32 |
| 121 – 130 | 58 |
| 131 – 140 | 9 |

**12.** 27 children are asked how many pets they own. How many more children have **3 or less** pets than have **6 or more** pets?

[1 2 3]  ▪ 15

Frequency distribution of pet ownership

| number of pets | number of children |
|----------------|---------------------|
| 0 – 1 | 4 |
| 2 – 3 | 15 |
| 4 – 5 | 4 |
| 6 – 7 | 2 |
| 8 – 9 | 2 |

**13.** How many more people watch television for 0 - 4 hours a day than for 5 - 8 hours a day?

▪ **32**

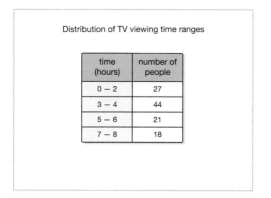

Distribution of TV viewing time ranges

| time (hours) | number of people |
|---|---|
| 0 − 2 | 27 |
| 3 − 4 | 44 |
| 5 − 6 | 21 |
| 7 − 8 | 18 |

**14.** Each pupil in class 7b is asked how many coloured pens they have in their pencil case. What range of coloured pens is the most common?

1/6  ▪ 0 - 5  ▪ **6 - 10**  ▪ 11 - 15  ▪ 16 - 20  ▪ 21 - 25  ▪ 26 - 30

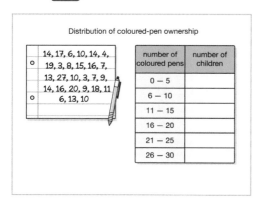

Distribution of coloured-pen ownership

14, 17, 6, 10, 14, 4, 19, 3, 8, 15, 16, 7, 13, 27, 10, 3, 7, 9, 14, 16, 20, 9, 18, 11 6, 13, 10

| number of coloured pens | number of children |
|---|---|
| 0 − 5 | |
| 6 − 10 | |
| 11 − 15 | |
| 16 − 20 | |
| 21 − 25 | |
| 26 − 30 | |

**15.** Some people are asked how many glasses of water they drink each day. Complete the frequency table, then sort the ranges in descending order (most common first).

▪ **0 - 5**  ▪ **11 - 15**  ▪ **6 - 10**  ▪ **more than 15**

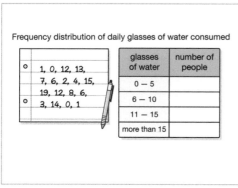

Frequency distribution of daily glasses of water consumed

1, 0, 12, 13, 7, 6, 2, 4, 15, 19, 12, 8, 6, 3, 14, 0, 1

| glasses of water | number of people |
|---|---|
| 0 − 5 | |
| 6 − 10 | |
| 11 − 15 | |
| more than 15 | |

**16.** A shop records how much money each customer spends to the nearest pound (£). What is the most common spending range?

1/6  ▪ 0 - 5  ▪ 6 - 10  ▪ 11 - 15  ▪ 16 - 20  ▪ **21 - 25**  ▪ 26 - 30

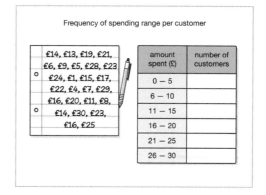

Frequency of spending range per customer

£14, £13, £19, £21, £6, £9, £5, £28, £23 £24, £1, £15, £17, £22, £4, £7, £29, £16, £20, £11, £8, £14, £30, £23, £16, £25

| amount spent (£) | number of customers |
|---|---|
| 0 − 5 | |
| 6 − 10 | |
| 11 − 15 | |
| 16 − 20 | |
| 21 − 25 | |
| 26 − 30 | |

**17.** How many flights cost £100 or more?

▪ **111**

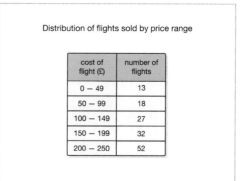

Distribution of flights sold by price range

| cost of flight (£) | number of flights |
|---|---|
| 0 − 49 | 13 |
| 50 − 99 | 18 |
| 100 − 149 | 27 |
| 150 − 199 | 32 |
| 200 − 250 | 52 |

**18.** Each pupil in class 8c is asked how many coloured pens they have in their pencil case. What range of coloured pens is the least common?

1/6  ▪ 0 - 5  ▪ 6 - 10  ▪ 11 - 15  ▪ 16 - 20  ▪ **21 - 25**  ▪ 26 - 30

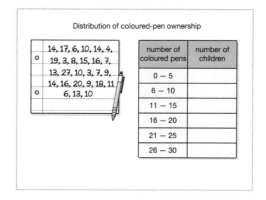

Distribution of coloured-pen ownership

14, 17, 6, 10, 14, 4, 19, 3, 8, 15, 16, 7, 13, 27, 10, 3, 7, 9, 14, 16, 20, 9, 18, 11 6, 13, 10

| number of coloured pens | number of children |
|---|---|
| 0 − 5 | |
| 6 − 10 | |
| 11 − 15 | |
| 16 − 20 | |
| 21 − 25 | |
| 26 − 30 | |

Level 2: *cont.*

**19.** Some pupils are asked about the countries they have visited. How many fewer children have been to Spain than Greece?

- 33

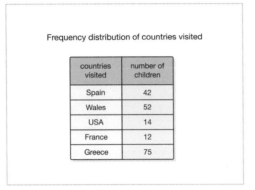

Frequency distribution of countries visited

| countries visited | number of children |
|---|---|
| Spain | 42 |
| Wales | 52 |
| USA | 14 |
| France | 12 |
| Greece | 75 |

**20.** 100 pupils are asked what their favourite sport is. How many more pupils choose rugby than snowboarding?

- 23

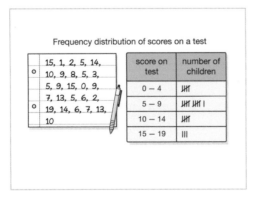

Frequency distribution of favourite sports

| sport | number of pupils |
|---|---|
| rugby | 25 |
| netball | 20 |
| football | 30 |
| horse riding | 23 |
| snowboarding | 2 |

**Level 3:** Reasoning - Reading frequency tables which require amendments.

✱ **Required:** 5/5    ✱ **Student Navigation:** on
✱ **Randomised:** off

**21.** Mr Jones has completed the frequency table using the data shown. Explain what mistake he has made.

Frequency distribution of scores on a test

15, 1, 2, 5, 14, 10, 9, 8, 5, 3, 5, 9, 15, 0, 9, 7, 13, 5, 6, 2, 19, 14, 6, 7, 13, 10

| score on test | number of children |
|---|---|
| 0 — 4 | JHI |
| 5 — 9 | JHI JHI I |
| 10 — 14 | JHI |
| 15 — 19 | III |

**22.** Jimmy asks 15 people how much they would pay to have their car washed. He makes a mistake when completing the table. What should the frequency for the range £2 – £3.99 be?

- 5

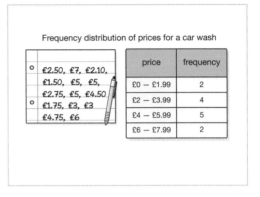

Frequency distribution of prices for a car wash

£2.50, £7, £2.10, £1.50, £5, £5, £2.75, £5, £4.50, £1.75, £3, £3, £4.75, £6

| price | frequency |
|---|---|
| £0 — £1.99 | 2 |
| £2 — £3.99 | 4 |
| £4 — £5.99 | 5 |
| £6 — £7.99 | 2 |

**23.** How many goals were scored in the league over the weekend?

- 129

Frequency of goals scored in the league

| goals scored | number of games | total: |
|---|---|---|
| 0 | 6 | |
| 1 | 4 | |
| 2 | 9 | |
| 3 | 11 | |
| 4 | 13 | |
| 5 | 2 | |
| 6 | 1 | |
| 7 | 0 | |

**24.** Isla surveys 20 people about their favourite snack, but makes a mistake when completing the table. What percentage choose apples?
*Don't include the % sign in your answer.*

- 20%  - 20

Frequency distribution of favourite snacks

| favourite snack | frequency | percentage |
|---|---|---|
| crisps | JHI | 25% |
| apple | IIII | 40% |
| orange | II | 10% |
| cereal bar | JHI IIII | 45% |

**25.** Jenny and Fred surveyed 30 children about what
time they go to bed. Fred says, "I think we should
make a line graph to show the results clearly." Is
Fred correct? Explain your answer.

a
b
c

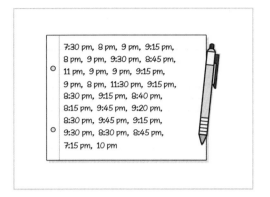

7:30 pm, 8 pm, 9 pm, 9:15 pm,
8 pm, 9 pm, 9:30 pm, 8:45 pm,
11 pm, 9 pm, 9 pm, 9:15 pm,
9 pm, 8 pm, 11:30 pm, 9:15 pm,
8:30 pm, 9:15 pm, 8:40 pm,
8:15 pm, 9:45 pm, 9:20 pm,
8:30 pm, 9:45 pm, 9:15 pm,
9:30 pm, 8:30 pm, 8:45 pm,
7:15 pm, 10 pm

**Level 4:**   Problem Solving - Complete frequency tables.

✳ **Required:** 5/5    ✳ **Student Navigation:** on
✳ **Randomised:** off

**26.** 102 children were asked to select their favourite
cartoon. The results for Tim & Jonny were
destroyed in the rain. How many more people
chose Tim & Jonny than The Limpsons?

1
2
3

▪ 23

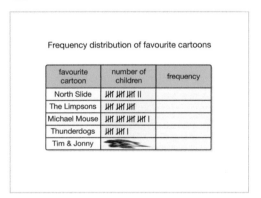

Frequency distribution of favourite cartoons

| favourite cartoon | number of children | frequency |
|---|---|---|
| North Slide | JHT JHT JHT II | |
| The Limpsons | JHT JHT JHT | |
| Michael Mouse | JHT JHT JHT JHT I | |
| Thunderdogs | JHT JHT I | |
| Tim & Jonny | | |

**27.** All children who attend the cinema are
accompanied by an adult. What is the mean
number of children per adult?

1
2
3

▪ 2.4

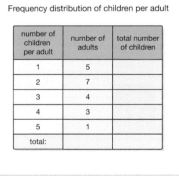

Frequency distribution of children per adult

| number of children per adult | number of adults | total number of children |
|---|---|---|
| 1 | 5 | |
| 2 | 7 | |
| 3 | 4 | |
| 4 | 3 | |
| 5 | 1 | |
| total: | | |

**28.** What percentage of darts players in the
tournament are left handed?
*Don't include the % sign in your answer.*

1
2
3

▪ 25

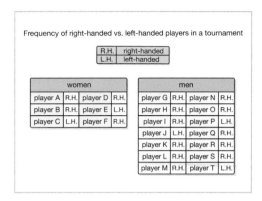

Frequency of right-handed vs. left-handed players in a tournament

| R.H. | right-handed |
|---|---|
| L.H. | left-handed |

| women | | | | men | | | |
|---|---|---|---|---|---|---|---|
| player A | R.H. | player D | R.H. | player G | R.H. | player N | R.H. |
| player B | R.H. | player E | L.H. | player H | R.H. | player O | R.H. |
| player C | L.H. | player F | R.H. | player I | R.H. | player P | L.H. |
| | | | | player J | L.H. | player Q | R.H. |
| | | | | player K | R.H. | player R | R.H. |
| | | | | player L | R.H. | player S | R.H. |
| | | | | player M | R.H. | player T | L.H. |

**29.** 512 children are asked which sport they prefer:
112 said hockey; 27 said snowboarding; 110 said
netball; 7 said darts and the rest said football.
What **percentage** prefer football?
*Don't include the % sign in your answer.*

1
2
3

▪ 50

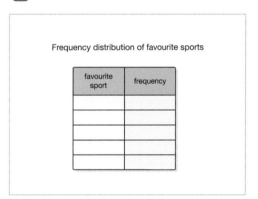

Frequency distribution of favourite sports

| favourite sport | frequency |
|---|---|
| | |
| | |
| | |
| | |

**30.** The owner of a cake stall keeps a list of how many
cakes each customer buys. What is the mean
number of cakes it sells to each customer?

1
2
3

▪ 3.25

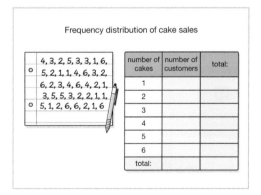

Frequency distribution of cake sales

4, 3, 2, 5, 3, 3, 1, 6,
5, 2, 1, 1, 4, 6, 3, 2,
6, 2, 3, 4, 6, 4, 2, 1,
3, 5, 5, 3, 2, 2, 1, 1,
5, 1, 2, 6, 6, 2, 1, 6

| number of cakes | number of customers | total: |
|---|---|---|
| 1 | | |
| 2 | | |
| 3 | | |
| 4 | | |
| 5 | | |
| 6 | | |
| total: | | |

# Complete, Read and Interpret Two-Way Tables

**Competency:** Construct and interpret appropriate tables, charts, and diagrams, including frequency tables, bar charts, pie charts, and pictograms for categorical data, and vertical line (or bar) charts for ungrouped and grouped numerical data.

**Quick Search Ref:** 10324

Correct: Correct.    Wrong: Incorrect, try again.    Open: Thank you.

**Level 1:** Understanding - Reading data and basic calculations in a two-way table.

❋ Required: 7/10    ❋ Student Navigation: on    ❋ Randomised: off

---

**1.** In the two-way table, which shape has more than 4 sides and angles which are all equal?

- irregular pentagon  ■ regular hexagon
1/5 ■ scalene triangle ■ square ■ equilateral triangle

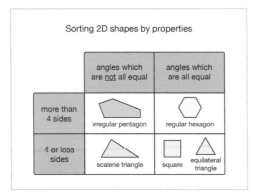

**2.** In Year 7, how many boys have a pet cat?

 ■ 12

Frequency of cat ownership in Year 7

|  | boys | girls |
|---|---|---|
| have a pet cat | 12 | 19 |
| do not have a pet cat | 15 | 13 |

**3.** How many points did the Ravens score in the javelin event?

 ■ 52

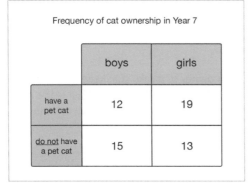

Frequency of points scored at an athletics competition

|  | relay | hurdles | javelin | shotput |
|---|---|---|---|---|
| Gulls | 100 | 80 | 12 | 100 |
| Eagles | 80 | 50 | 20 | 76 |
| Ravens | 90 | 100 | 52 | 92 |
| Hawks | 70 | 70 | 98 | 18 |
| Kestrels | 50 | 90 | 78 | 12 |

**4.** 13 children go to Harry's birthday party. How many girls are aged **11 or over**?

 ■ 5

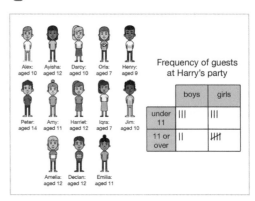

Frequency of guests at Harry's party

|  | boys | girls |
|---|---|---|
| under 11 | III | III |
| 11 or over | II | IIII |

**5.** According the table, how many people in the office drink tea with sugar?

 ■ 6

Frequency of hot drink preferences

|  | no sugar | with sugar | total: |
|---|---|---|---|
| tea | ? | ? | ? |
| coffee | 9 | 3 | 12 |
| total: | 18 | 9 | 27 |

**6.** How many season ticket holders are male?

 ■ 31

Frequency of season ticket holders by age and gender

|  | male | female | total: |
|---|---|---|---|
| under 16 |  | 25 | 42 |
| 16 or over |  | 32 | 46 |
| total: | ? |  |  |

**7.** How many children in the class have brown eyes?

  ▪ **11**

Distribution of eye colour by gender

|         | boys | girls | total: |
|---------|------|-------|--------|
| green   | 7    | 8     | 15     |
| blue    | 5    | 4     | 9      |
| brown   | 4    | 7     | ?      |
| total:  | 16   | 19    | 35     |

**8.** How many people visited Karl's Cafe on Tuesday?

  ▪ **22**

Frequency of visitors to shops in a town

|                | Mon | Tues | Wed | Thurs | Fri | Sat |
|----------------|-----|------|-----|-------|-----|-----|
| Roy's Racing   | 43  | 92   | 82  | 15    | 71  | 19  |
| Pound Planet   | 789 | 363  | 331 | 149   | 364 | 519 |
| Super-shoppa   | 127 | 238  | 625 | 128   | 135 | 456 |
| Karl's Cafe    | 72  | 22   | 54  | 21    | 53  | 89  |
| The Meat House | 82  | 98   | 61  | 87    | 35  | 0   |

**9.** How many children have a pet cat in total?

  ▪ **31**

Frequency distribution of cat ownership

|                        | boys | girls | total: |
|------------------------|------|-------|--------|
| have a pet cat         | 12   | 19    | ?      |
| do not have a pet cat  | 15   | 13    | 28     |
| total:                 | 27   | 32    | 59     |

**10.** A survey asked people whether they prefer white chocolate, milk chocolate or dark chocolate. How many females took part in the survey?

▪ **27**

Frequency distribution of chocolate preferences by gender

|         | white | milk | dark | total: |
|---------|-------|------|------|--------|
| females | 7     | 12   | 8    | ?      |
| males   | 10    | 21   | 5    | 36     |
| total:  | 17    | 33   | 13   | 63     |

**11.** How many pupils **in total** have packed lunches?

 ▪ **112**

Distribution of school dinner preferences

|                | Year 7 | Year 8 | Year 9 | Year 10 | Year 11 | total: |
|----------------|--------|--------|--------|---------|---------|--------|
| packed lunches | 32     | 20     | 18     | 30      | 12      | ?      |
| school dinners | 18     | 46     | 22     | 19      | 39      | 144    |
| total:         | 50     | 66     | 40     | 49      | 51      | 256    |

**12.** How many more children watch television for **more than 3 hours** a day than for **2 - 3 hours** a day?

 ▪ **8**

Distribution of time spent watching televison in a day

|                  | age 11 | age 12 | age 13 | total: |
|------------------|--------|--------|--------|--------|
| less than 2 hours| 15     | 12     | 19     |        |
| 2 – 3 hours      | 10     | 46     | 22     |        |
| more than 3 hours| 18     | 46     | 22     |        |
| total:           |        |        |        |        |

**13.**  How many pupils in Year 8 said that geography was their favourite subject?

 ■ 20

Distribution of favourite subjects

|  | Year 7 | Year 8 | Year 9 | total: |
|---|---|---|---|---|
| English | 50 | ? | 10 | 75 |
| geography | 20 | ? | ? | ? |
| maths | 20 | ? | 35 | 110 |
| total: | 90 | 90 | 90 | 270 |

**14.** 91 Year 7 pupils were asked to choose their favourite sport. Arrange the categories in descending order (highest first).

■ girls who play football ■ boys who play rugby
■ boys who play hockey ■ girls who play rugby

Distribution of sport preferences by gender

|  | football | rugby | hockey | total: |
|---|---|---|---|---|
| boys | 33 |  |  | 52 |
| girls |  |  | 19 |  |
| total: |  | 13 | 28 | 91 |

**15.** A shop records how much of each product they sell each week. How many oranges are sold in total?

 ■ 57

Distribution of products sold over 3 weeks

|  | week 1 | week 2 | week 3 | total: |
|---|---|---|---|---|
| apples | 21 | 13 | 25 | 59 |
| oranges | ? | 32 | ? | ? |
| bananas | 15 | 22 | 15 | ? |
| potatoes | 19 | 16 | 20 | ? |
| total: | 65 | 83 | 75 | ? |

**16.** A geography class is asked whether they have been to Spain, Greece or France. How many more children have been to Spain than Greece?

 ■ 33

Frequency distribution of countries visited by gender

|  | Spain | Greece | France | total: |
|---|---|---|---|---|
| boys |  | 27 |  | 80 |
| girls |  |  | 7 | 58 |
| total: |  | 42 | 21 |  |

**17.** How many more women entered the swimming competition than men?

■ 13

Distribution of swimming competition entrants by gender

|  | front crawl | breast stroke | back stroke | butterfly | total: |
|---|---|---|---|---|---|
| male |  | 10 |  | 12 |  |
| female | 21 |  | 17 |  |  |
| total: | 39 |  | 36 | 31 | 131 |

**18.** How many Easywings flights cost more than £300?

 ■ 111

Frequency distribution of flight cost

|  | Bryanair | Easywings | Jet4 | total: |
|---|---|---|---|---|
| less than £100 | ? | 18 | ? | 105 |
| £100 – £300 | 51 | 150 | 102 | 303 |
| more than £300 | 249 | ? | 99 | ? |
| total: | 312 | 279 | 276 | 867 |

Ref:10324    Complete, Read and Interpret Two-Way ...

**19.** How many more Year 7 pupils than Year 11 pupils have school dinners?

1
2
3   ▪ 23

Distribution of school dinner preferences

|  | Year 7 | Year 8 | Year 9 | Year 10 | Year 11 | total: |
|---|---|---|---|---|---|---|
| school dinners | ? | ? | ? | ? | ? | 381 |
| packed lunches | 35 | 52 | 45 | 36 | 72 | 240 |
| total: | 121 | 137 | 117 | 111 | 135 | 621 |

**20.** 72 children are asked what pets they own. Arrange the categories in ascending order (smallest first).

↑
↓

▪ boys with a rabbit  ▪ boys with a dog
▪ girls with a dog  ▪ girls with a cat

Frequency distribution of pets owned by gender

|  | cat | dog | rabbit | total: |
|---|---|---|---|---|
| boys | 7 |  |  | 34 |
| girls |  |  | 5 |  |
| total: |  | 31 | 17 | 72 |

✸ **Required:** 5/5   ✸ **Student Navigation:** on
✸ **Randomised:** off

**21.** Which shape is in the wrong place in the two-way table?

▪ 1 ▪ 2 ▪ 3 ▪ 4 ▪ 5 ▪ 6 ▪ 7

1/7

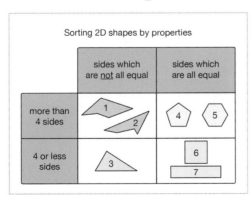

Sorting 2D shapes by properties

**22.** Billy has completed the two-way table using the data shown. Explain what mistake he has made.

a
b
c

Frequency distribution of shapes by colour

|  | ○ | □ | △ | total: |
|---|---|---|---|---|
| red | 2 | 4 | 1 | 7 |
| blue | 3 | 2 | 3 | 8 |
| green | 1 | 1 | 3 | 5 |
| total: | 6 | 7 | 7 |  |

**23.** Temporary traffic lights mean every bus journey is taking 18 minutes longer. What is the first bus that will arrive at Finchley after 10:00?

1/5   ▪ bus A ▪ bus B ▪ bus C ▪ bus D ▪ bus E

Bus times from Keenal to Brent

|  | bus A | bus B | bus C | bus D | bus E |
|---|---|---|---|---|---|
| Keenal | 07:42 | 07:59 | 08:37 | 09:06 | 09:31 |
| Tintou | 08:14 | 08:31 | 09:04 | 09:35 | 10:03 |
| Finchley | 08:31 | 08:50 | 09:25 | 09:52 | 10:21 |
| Brent | 08:47 | 09:05 | 09:39 | 10:01 | 10:33 |

**24.** Sam lives in Soweray and works in Hepton. It takes her 12 minutes to walk from home to Soweray train station and another 15 minutes to walk from Hepton train station to work. What is the latest time Sam can leave her house if she starts work at 11:30?

a
b
c

▪ 10:24

Train times from Soweray to Hepton

| Soweray | 10:15 | 10:31 | 10:36 | 10:40 | 10:48 |
|---|---|---|---|---|---|
| Trafax | 10:33 | 10:45 | 10:51 | 10:56 | 11:02 |
| Foulder | 10:46 | 10:53 | 10:59 | 11:05 | 11:13 |
| Hepton | 10:59 | 11:05 | 11:12 | 11:18 | 11:24 |

**Level 3: cont.**

**25.** According to the table, how many children in Year 7 have two siblings?

■ 91

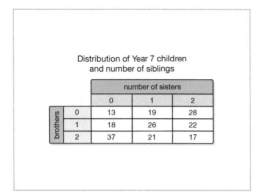

Distribution of Year 7 children
and number of siblings

| | | number of sisters | | |
|---|---|---|---|---|
| | | 0 | 1 | 2 |
| brothers | 0 | 13 | 19 | 28 |
| | 1 | 18 | 26 | 22 |
| | 2 | 37 | 21 | 17 |

**Level 4:** Problem Solving - Complete two-way tables.

✳ **Required:** 5/5    ✳ **Student Navigation:** on
✳ **Randomised:** off

**26.** 35 children are asked whether they would prefer to go to a science museum or theme park. How many boys want to go to the science museum?
11 children said science museum.
12 girls said theme park.
There are 17 boys in total.

■ 5

Distribution of school trip preferences by gender

| | boys | girls | total: |
|---|---|---|---|
| science museum | | | |
| theme park | | | |
| total: | | | |

**27.** A farmer counts how many animals he sees in one of his fields each day. How many sheep did he see on day 2?

■ 17

Frequency of animal sightings in a farmers field

| | day 1 | day 2 | day 3 | day 4 | day 5 | total: |
|---|---|---|---|---|---|---|
| cows | 10 | 11 | 9 | 12 | 7 | ? |
| sheep | ? | ? | ? | 15 | ? | 75 |
| horses | 14 | 10 | 12 | 10 | 14 | ? |
| pigs | 5 | ? | 5 | 4 | 5 | 24 |
| total: | 47 | ? | 39 | ? | 38 | 208 |

**28.** What percentage of darts players in the tournament are male **and** left-handed?
*Don't include the % sign in your answer.*

■ 34

Frequency distribution of right-handed and left-handed
players in a darts tournament

| | male | female | total: |
|---|---|---|---|
| right-handed | ? | 18 | 28 |
| left-handed | ? | ? | ? |
| total: | ? | 23 | 50 |

**29.** 418 children are asked which sport they would prefer. If 186 of the children asked were girls and 172 children said rugby, how many girls preferred rugby?
44 boys said hockey.
65 boys said tennis.
78 girls said tennis.

■ 49

Distribution of sport preferences by gender

| | hockey | tennis | rugby | total: |
|---|---|---|---|---|
| boys | | | | |
| girls | | | | |
| total: | | | | |

**30.** A cake shop records the different types of cake sold each day. What is the mean number of cakes sold each day?

■ 231

Frequency distribution of cakes sold in a week

| | Mon | Tue | Wed | Thurs | Fri | Sat | Sun | total: |
|---|---|---|---|---|---|---|---|---|
| scones | 71 | 212 | ? | 67 | 33 | 19 | 54 | 643 |
| iced fingers | 270 | 105 | 35 | 45 | 0 | 52 | 13 | ? |
| cupcakes | 29 | 100 | 130 | ? | 74 | 7 | 46 | 514 |
| total: | ? | ? | ? | ? | ? | ? | ? | ? |

# Interpret Bar Charts, Pictograms and Line Graphs

**Competency:** Construct and interpret appropriate tables, charts, and diagrams, including frequency tables, bar charts, pie charts, and pictograms for categorical data, and vertical line (or bar) charts for ungrouped and grouped numerical data.

**Quick Search Ref:** 10323

**Correct:** Correct.    **Wrong:** Incorrect, try again.    **Open:** Thank you.

**Level 1:** Understanding - Interpreting data in a line graph, bar chart or pictogram.

❊ **Required:** 7/10    ❊ **Student Navigation:** on    ❊ **Randomised:** off

1. Jamie records his heart rate every hour for 12 hours. Which graph or chart would best represent this data?

   1/4  ■ bar chart ■ pie chart ■ line graph ■ pictogram

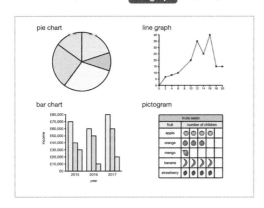

2. Which set of data would you present in a bar chart?

   ■ Change in money in a bank account over a year.
   ■ Number of boys and girls in different classes.
   1/4
   ■ Change of temperature in a country in a month.
   ■ Change of heart rate during a day.

3. Jason records how long it takes to get from Luffton to Hepton using different methods of transport. Which bar chart correctly represents this data?

   1/3  ■ bar chart (i) ■ bar chart (ii) ■ bar chart (iii)

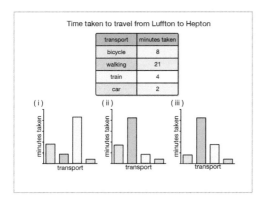

4. Andy and his friends count their football cards. Who has fewer than 30 cards?

   1/5  ■ Andy ■ Gary ■ Amanda ■ Pete ■ Sarah

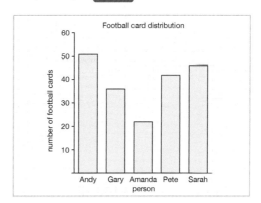

5. What is the average temperature in Westhall in the warmest month?
   *Don't include the units in your answer.*

   ■ 27

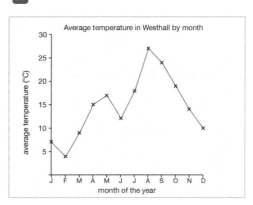

6. How many museum visitors arrive at 3 p.m. or later?

   ■ 450

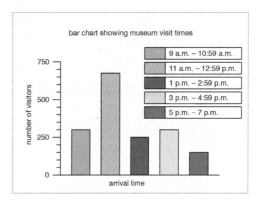

**Level 1:** *cont.*

**7.** According to the chart, how many DVDs did LMV Music sell?

■ 34

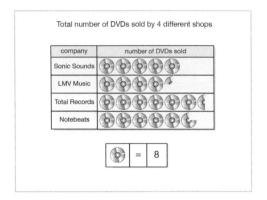

**8.** Which member of the Smith family is 15 years old?

■ Dave ■ Ben ■ Ellie ■ Sam ■ Rachel

1/5

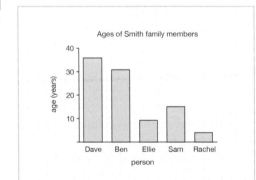

**9.** 100 people were asked what type of car they drive and the data was recorded in a table. Select the bar chart which represents the data.

1/3 ■ bar chart (i) ■ bar chart (ii) ■ bar chart (iii)

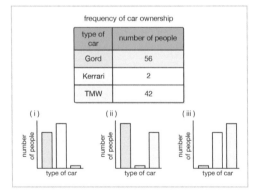

**10.** Which set of data would you present in a line graph?

■ After-school club membership.

1/4 ■ The times taken for athletes to run 100 metres.

■ Number of pets in a household.

■ The heights of pupils in a class.

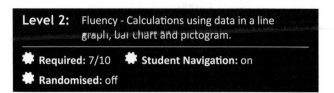

✱ **Required:** 7/10   ✱ **Student Navigation:** on
✱ **Randomised:** off

**11.** If Ana sold 45 apples on her market stall on Wednesday, what is the missing value in the key to the pictogram?

■ 10

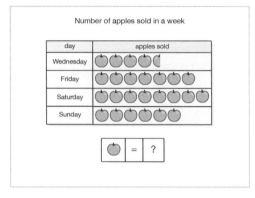

**12.** In 2010, how much greater was the population in Las Farmas than Rockvale?

■ 75000 ■ 75,000

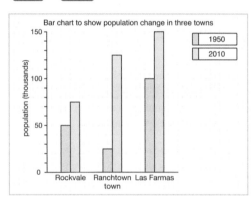

**13.** How much profit did Jango's Jewellery make from Friday to Sunday?
*Include the £ sign in your answer.*

■ £230.00 ■ £230

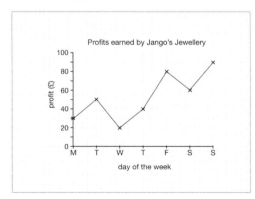

**14.** How many more hours sleep did Sarah get on Tuesday than on Saturday?

- **2**

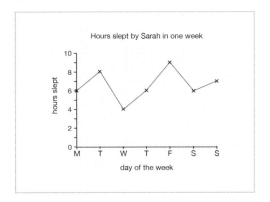

**15.** The number of customers at Regal Bakery are totalled for two day periods and this information is shown in the pictogram. Select the line graph which represents this data.

1/3

- **line graph (i)** ▪ **line graph (ii)** ▪ **line graph (iii)**

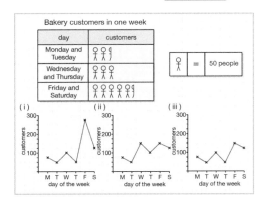

**16.** The bar chart shows the number of students applying for a university in the months between January and June. How many **female** students applied in April?

- **175**

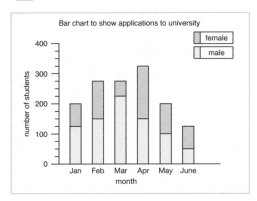

**17.** If a pizza shop sold 130 pizzas on Friday, what number completes the key to the pictogram?

- **40**

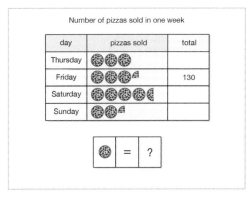

**18.** A shop compares the amount of orange juice and apple juice it sells on each day in one week. Arrange the quantities sold on each day in descending order (largest quantity first).

- Apple juice on Monday ▪ Apple juice on Thursday
- Orange juice on Monday ▪ Orange juice on Wednesday
- Apple juice on Tuesday ▪ Apple juice on Wednesday

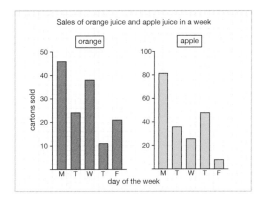

**19.** The bar chart shows the number of students applying for a university in the months between January and June. How many more **male** students applied in March than in June?

- **175**

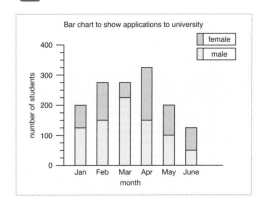

**20.** The pictogram shows the number of Year 7 pupils who travel to school by bus, car and on foot. How many more children walk to school than travel by bus?

■ 40

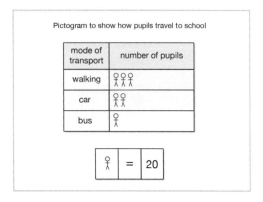

Pictogram to show how pupils travel to school

**23.** The pictogram shows the number of boys and girls who attend two different schools. Aliysa says "I can't tell how many girls attend Yowton High." Jane says, "15 girls attend Yowton High." Who is correct? Explain your answer.

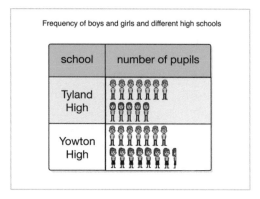

**Level 3:** Reasoning - Pictograms, bar charts and line graphs with missing data.

✻ **Required:** 5/5    ✻ **Student Navigation:** on
✻ **Randomised:** off

**21.** According to the graph, how many minutes did Tim ride his bike for during his bike ride?

■ 120

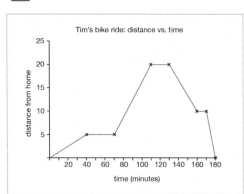

**24.** The bar chart is incomplete. If a golf club has 241 members, how many members are aged over 50?

■ 45

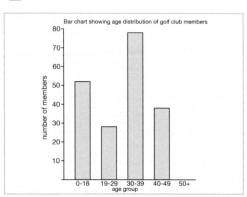

**22.** Children from Bopton High and Rivdale High are asked if they play a musical instrument. Sort the responses for each option in descending order (largest first).

■ Bopton High - Yes   ■ Bopton High - No
■ Rivdale High - Yes   ■ Rivdale High - No
■ Rivdale High - No comment
■ Bopton High - No comment

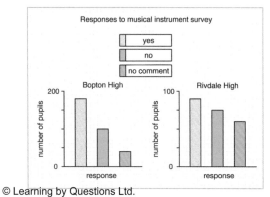

**25.** A family go to the beach for the day and return home in the evening. They stop once for a break on the way there but **not** on the way back. Which line graph shows their journey? Explain your choice.

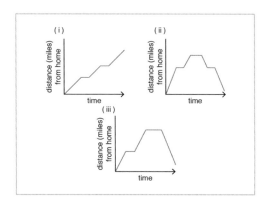

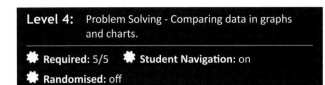

**Level 4:** Problem Solving - Comparing data in graphs and charts.

✳ **Required:** 5/5    ✳ **Student Navigation:** on
✳ **Randomised:** off

**26.** According to the pictogram, how many ice-creams were sold in one week?

  ▪ 190

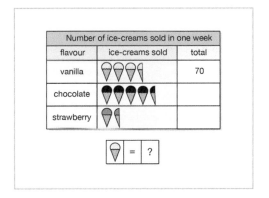

**27.** How many more people register with the three mobile phone companies in year 2 than in year 1?

  ▪ 40,000  ▪ 40000

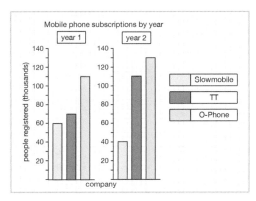

**28.** Use the data shown on the graphs to work out how many more siblings of children in Class 6 there are, than siblings of children in Class 5.

▪ 11

**29.** How many minutes faster was the fastest boy in the cross-country race than the fastest girl?

▪ 4

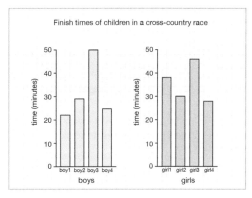

**30.** Calculate the mean number of houses on a street based on the data in the pictogram.

▪ 20

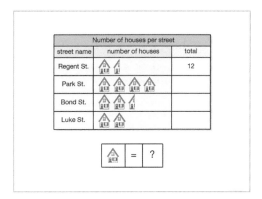

# Interpret bar charts, pie charts, pictograms and line graphs

**Competency:** Construct and interpret appropriate tables, charts and diagrams, including bar charts, pie charts and pictograms.

**Quick Search Ref:** 10074

**Correct:** Correct.    **Wrong:** Incorrect, try again.    **Open:** Thank you.

**Level 1:** Understanding - Interpreting data in a line graph, pie chart, bar chart or pictogram.

✱ **Required:** 7/10    ✱ **Student Navigation:** on    ✱ **Randomised:** off

**1.** Jamie records his heart rate every hour for 12 hours. Which graph or chart would best represent this data?

1/4    ▪ **bar chart** ▪ **pie chart** ▪ **line graph** ▪ **pictogram**

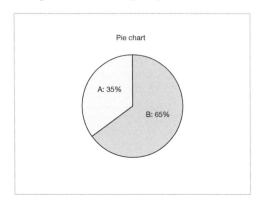

**2.** Which set of data would you present in a pie chart?

▪ **Hours spent cooking each day.**
1/4    ▪ **Number of boys and girls in a class.**
▪ **Change of temperature in a country each month.**
▪ **Change of heart rate during a day.**

**3.** Jason records how long it takes to get from Luffton to Hepton using different methods of transport. Which bar chart correctly represents this data?

1/3    ▪ **bar chart (i)** ▪ **bar chart (ii)** ▪ **bar chart (iii)**

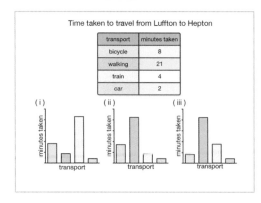

**4.** Andy and his friends count their football cards. Who has fewer than 30 cards?

▪ **Andy** ▪ **Gary** ▪ **Amanda** ▪ **Pete** ▪ **Sarah**
1/5

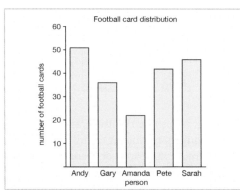

**5.** What is the average temperature in Westhall in the warmest month?
*Don't include the units in your answer.*

▪ **27**

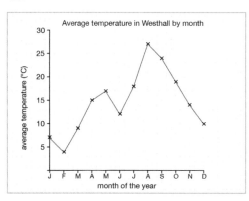

**Level 1:** *cont.*

**6.** What percentage of museum visitors arrive at 2 p.m. or after?
*Don't include the % sign in your answer.*

- **65**

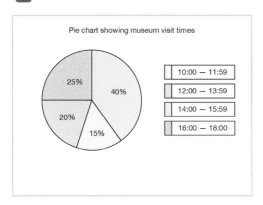

Pie chart showing museum visit times

25%
40%
20%
15%

| 10:00 — 11:59 |
| 12:00 — 13:59 |
| 14:00 — 15:59 |
| 16:00 — 18:00 |

**7.** According to the chart, how many DVDs did LMV Music sell?

- **34**

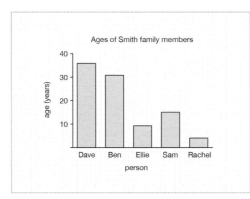

Total number of DVDs sold by 4 different shops

| company | number of DVDs sold |
|---|---|
| Sonic Sounds | |
| LMV Music | |
| Total Records | |
| Notebeats | |

| | = | 8 |

**8.** Which member of the Smith family is 15 years old?

- Dave ■ Ben ■ Ellie ■ **Sam** ■ Rachel

1/5

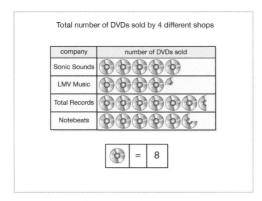

Ages of Smith family members

age (years)
40
30
20
10

Dave  Ben  Ellie  Sam  Rachel
person

**9.** 100 people were asked what type of car they drive and the data was recorded in a table. Select the pie chart which represents the data.

1/3   ■ pie chart (i)   ■ **pie chart (ii)**   ■ pie chart (iii)

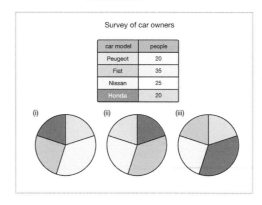

Survey of car owners

| car model | people |
|---|---|
| Peugeot | 20 |
| Fiat | 35 |
| Nissan | 25 |
| Honda | 20 |

(i)        (ii)        (iii)

**10.** Which set of data would you present in a line graph?

- After-school club membership.
1/4 ■ **The times taken for athletes to run 100 metres.**
- Number of pets in a household.
- The heights of pupils in a class.

**Level 2:** Fluency - Calculations using data in a line graph, pie chart, bar chart and pictogram.

✱ **Required:** 7/10   ✱ **Student Navigation:** on
✱ **Randomised:** off

**11.** If Ana sold 45 apples on her market stall on Wednesday, what is the missing value in the key to the pictogram?

- **10**

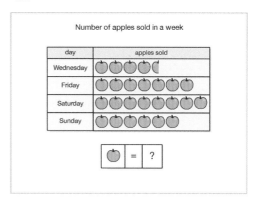

Number of apples sold in a week

| day | apples sold |
|---|---|
| Wednesday | |
| Friday | |
| Saturday | |
| Sunday | |

| | = | ? |

**12.** The total population of Rockvale Valley is 49,000.
How many people live in Backdown if an identical
number of people live in Backdown and Hayton?

a b c

■ 5,000 ■ 5000

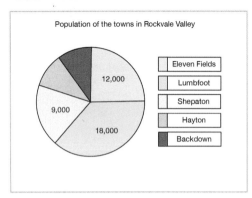

**13.** How much profit did Jango's Jewellery make from
Friday to Sunday?
*Include the £ sign in your answer.*

a b c

■ £230 ■ £230.00

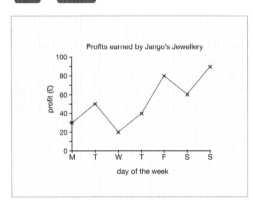

**14.** How many more hours sleep did Sarah get on
Tuesday than on Saturday?

1 2 3

■ 2

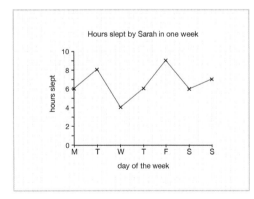

**15.** The number of customers at Regal Bakery are
totalled for two day periods and this information is
shown in the pictogram. Select the line graph
which represents this data.

1/3

■ (i) ■ (ii) ■ (iii)

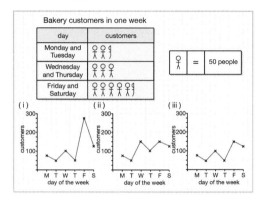

**16.** The pie chart shows the number of students
applying for university in the months between
January and June. Which bar chart represents this
data?

1/2

■ A ■ B

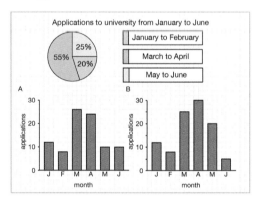

**17.** If the refreshments for a council meeting cost £20
and tea and coffee cost the same, how much was
spent on sugar?
*Include the £ sign in your answer.*

a b c

■ £2.25

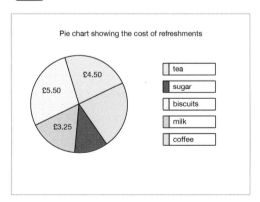

**18.** If a pizza shop sold 130 pizzas on Friday, what number completes the key to the pictogram?

1 2 3  ▪ 40

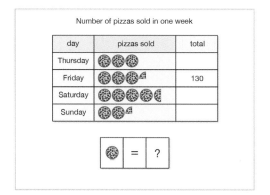

**19.** A shop compares the amount of orange juice and apple juice it sells on each day in one week. Arrange the quantities sold on each day in descending order (largest quantity first).

↑↓

- ▪ Apple juice on Monday  ▪ Apple juice on Thursday
- ▪ Orange juice on Monday  ▪ Orange juice on Wednesday
- ▪ Apple juice on Tuesday  ▪ Apple juice on Wednesday

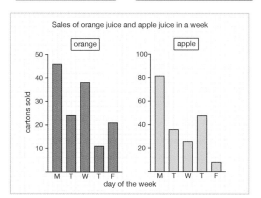

**20.** The pictogram shows the number of Year 7 pupils who travel to school by bus, car and on foot. How many more children walk to school than travel by bus?

1 2 3

▪ 40

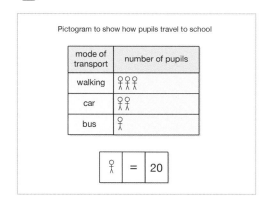

**Level 3:** Reasoning - Pictograms, bar charts, pie charts and line graphs with missing data.

✱ Required: 5/5  ✱ Student Navigation: on
✱ Randomised: off

**21.** According to the graph, how many minutes did Tim ride his bike for during his bike ride?

1 2 3  ▪ 120

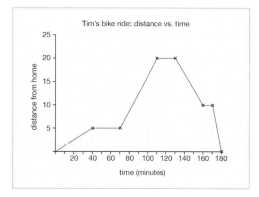

**22.** 200 children from Bopton High and 100 children from Rivdale High are asked if they play a musical instrument. Sort the responses for each option in descending order (highest first).

↑↓

- ▪ Bopton High - Yes  ▪ Bopton High - No
- ▪ Rivdale High - Yes  ▪ Rivdale High - No
- ▪ Rivdale High - No comment
- ▪ Bopton High - No comment

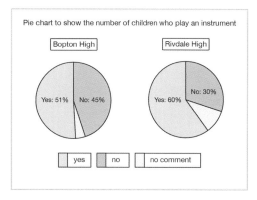

**23.** The pie charts show the number of boys and girls who attend two different schools. Which school has more boys? Explain your answer.

a b c

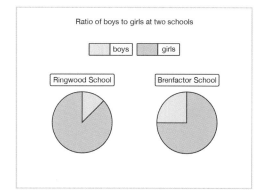

Level 3: *cont.*

**24.** The bar chart is incomplete. If a golf club has 241 members, how many members are aged over 50?

**1 2 3**   ▪ **45**

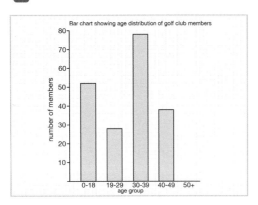

**25.** A survey asked a number of people to name their favourite holiday destination. If 17 people said Mexico, how many people were surveyed in total?

**1 2 3**

▪ **68**

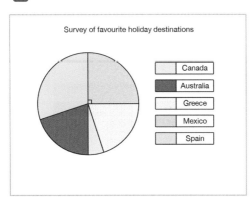

**Level 4:**   Problem Solving - Comparing data on more than one graph or chart.

✻ **Required:** 5/5   ✻ **Student Navigation:** on
✻ **Randomised:** off

**26.** According to the pictogram, how many ice-creams were sold in one week?

**1 2 3**   ▪ **190**

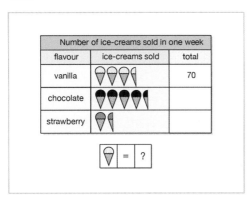

**27.** How many more people register with the three mobile phone companies in year 2 than in year 1?

**a b c**   ▪ **40,000**  ▪ **40000**

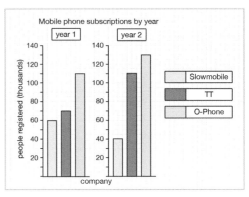

**28.** Use the data shown on the graphs to work out how many more siblings of children in class 6 there are, than siblings of children in class 5.

**1 2 3**

▪ **11**

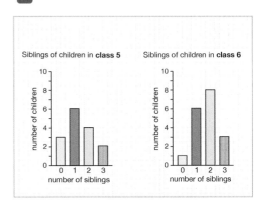

**29.** How many minutes faster was the fastest boy in the cross-country race than the fastest girl?

**1 2 3**   ▪ **6**

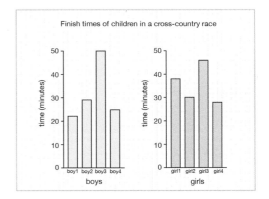

**30.** Calculate the mean number of houses on a street based on the data in the pictogram.

■ 20

| Number of houses per street | | |
|---|---|---|
| street name | number of houses | total |
| Regent St. | 🏠 🏠 | 12 |
| Park St. | 🏠 🏠 🏠 🏠 | |
| Bond St. | 🏠 🏠 🏠 | |
| Luke St. | 🏠 🏠 | |

| 🏠 | = | ? |
|---|---|---|

## Interpret Pie Charts

**Competency:** Construct and interpret appropriate tables, charts, and diagrams, including frequency tables, bar charts, pie charts, and pictograms for categorical data, and vertical line (or bar) charts for ungrouped and grouped numerical data.

**Quick Search Ref:** 10322

Correct: Correct.     Wrong: Incorrect, try again.     Open: Thank you.

**Level 1:** Understanding - Interpreting data in a pie chart.

⚙ **Required:** 7/10      ⚙ **Student Navigation:** on      ⚙ **Randomised:** off

---

**1.** Which set of data would you present in a pie chart?

- Change of water depth in a river during a day.
- Number of boys and girls in a class.
- Change of temperature in a country each month.
- Change of heart rate during a day.

1/4

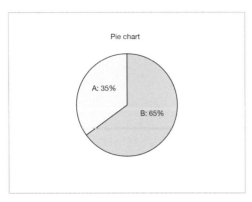

**2.** Jason records how long it takes to get from Luffton to Hepton using different methods of transport. Which pie chart correctly represents this data?

1/3   ▪ pie chart (i)  ▪ pie chart (ii)  ▪ pie chart (iii)

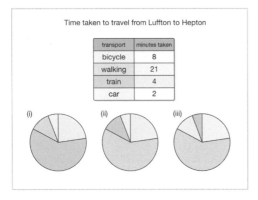

**3.** Andy and his friends have 130 football cards altogether. Who has the fewest cards?

▪ Andy ▪ Gary ▪ Amanda ▪ Pete

1/4

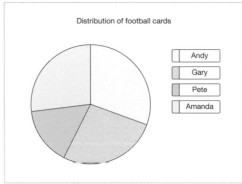

**4.** Arrange the TV channels in order, from the most popular to the least popular.

▪ Channel 72 ▪ LBC 1 ▪ The Exploring Channel
▪ DTV ▪ LBC 2

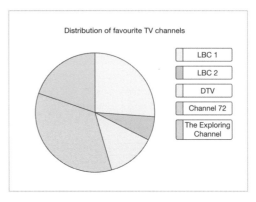

**5.** 100 pupils are asked to choose their favourite subject. How many choose science?

1
2
3   ▪ 50

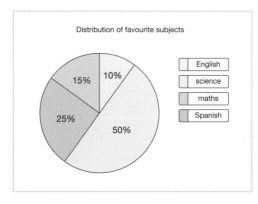

---

**6.** What percentage of museum visitors arrive at 2 p.m. or after?

1
2
3

*Don't include the % sign in your answer.*

■ 65

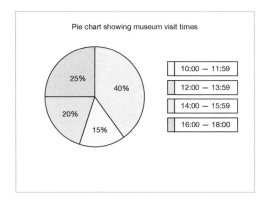

**7.** LMV sells 200 products in a day. How many CDs are sold?

1
2
3

■ 50

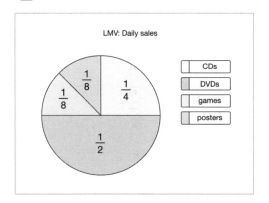

**8.** 100 people were asked what type of car they drive and the data was recorded in a table. Select the pie chart which represents the data.

☐
☒
☐

1/3　■ pie chart (i)　■ pie chart (ii)　■ pie chart (iii)

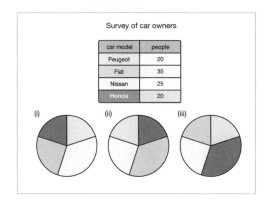

**9.** 100 pupils are asked to choose their favourite subject. How many choose Spanish?

1
2
3

■ 25

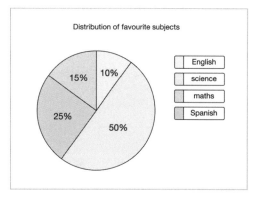

**10.** What percentage of museum visitors arrive **before** 2 p.m.?

1
2
3

*Don't include the % sign in your answer.*

■ 35

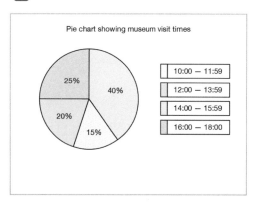

**Level 2:** Fluency - Calculations using data in a pie chart.
___
✿ **Required:** 7/10　✿ **Student Navigation:** on
✿ **Randomised:** off

**11.** A survey asks people to choose their favourite restaurant. If 500 choose Big L's, how many choose The Pack and Mutton?

1
2
3

■ 250

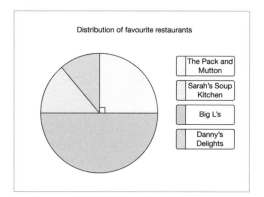

**12.** The total population of Rockvale Valley is 49,000.
How many people live in Backdown if the same
number of people live in Backdown and Hayton?

a
b
c

▪ 5,000 ▪ 5000

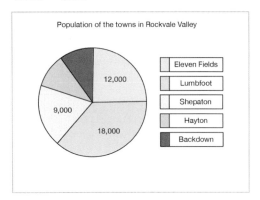

Population of the towns in Rockvale Valley

| | |
|---|---|
| 12,000 | Eleven Fields |
| 9,000 | Lumbfoot |
| 18,000 | Shepaton |
| | Hayton |
| | Backdown |

**13.** Jango's Jewellery records the items it sells in a
week. How many more bracelets are sold than
rings?

a
b
c

▪ 1854 ▪ 1,854

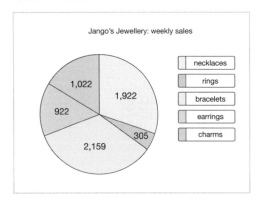

Jango's Jewellery: weekly sales

| | |
|---|---|
| 1,022 | necklaces |
| 1,922 | rings |
| 922 | bracelets |
| 2,159 | earrings |
| 305 | charms |

**14.** 600 children are asked to choose their favourite
animal. How many choose dolphins?

1
2
3

▪ 150

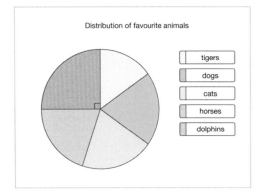

Distribution of favourite animals

| |
|---|
| tigers |
| dogs |
| cats |
| horses |
| dolphins |

**15.** The products sold in a bakery in a week are shown
on a pie chart. 300 cupcakes were sold. The
number of cupcakes and teacakes are equal. How
many steak pies were sold?

a
b
c

▪ 600

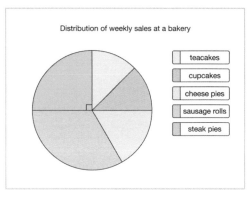

Distribution of weekly sales at a bakery

| |
|---|
| teacakes |
| cupcakes |
| cheese pies |
| sausage rolls |
| steak pies |

**16.** The pie chart shows the percentage of students
applying for university in the months between
January and June. What percentage of students
applied in January and February?
*Don't include the % sign in your answer.*

1
2
3

▪ 20

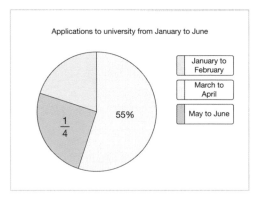

Applications to university from January to June

| | |
|---|---|
| 55% | January to February |
| $\frac{1}{4}$ | March to April |
| | May to June |

**17.** If the refreshments for a council meeting cost £20
and tea and coffee cost the same, how much was
spent on sugar?
*Include the £ sign in your answer.*

a
b
c

▪ £2.25

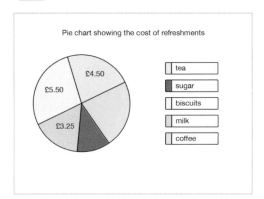

Pie chart showing the cost of refreshments

| | |
|---|---|
| £4.50 | tea |
| £5.50 | sugar |
| £3.25 | biscuits |
| | milk |
| | coffee |

**Level 2: *cont.***

**18.** 800 children are asked to choose their favourite animal. How many choose dogs?

**1**
**2**
**3**
 ▪ 600

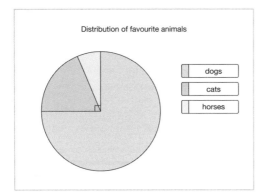

Distribution of favourite animals

dogs
cats
horses

**19.** Jango's Jewellery records the items it sells in a week. How many more necklaces are sold than earrings?

**a**
**b**
**c**
 ▪ 1000 ▪ 1,000

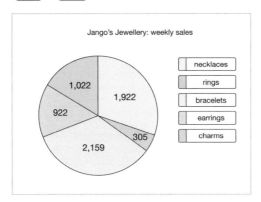

Jango's Jewellery: weekly sales

necklaces
rings
bracelets
earrings
charms

1,022   1,922
922
305
2,159

**20.** The products sold in a bakery in a week are shown on a pie chart. 500 steak pies were sold. How many products were sold in **total**?

**a**
**b**
**c**
 ▪ 2,000 ▪ 2000

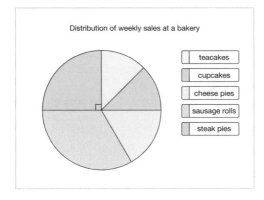

Distribution of weekly sales at a bakery

teacakes
cupcakes
cheese pies
sausage rolls
steak pies

---

**Level 3:** Reasoning - Pie charts with missing and incorrect data.

✳ **Required:** 5/5   ✳ **Student Navigation:** on

✳ **Randomised:** off

**21.** According to the pie charts, select **all** statements which are **correct**.

☐
☒
☐
3/6

 ▪ **One quarter of girls walk to school.**
 ▪ One quarter of boys walk to school.
 ▪ More than half of girls walk to school.
 ▪ **25 boys walk to school.** ▪ 25 girls take the bus to school.
 ▪ **In total, exactly half of the children walk to school.**

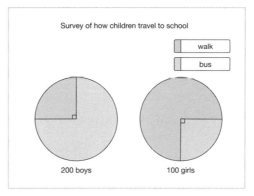

Survey of how children travel to school

walk
bus

200 boys        100 girls

**22.** 200 children from Bopton High and 100 children from Rivdale High are asked if they play a musical instrument. Sort the responses for each option in descending order (highest first).

↑
↓

 ▪ Bopton High - Yes   ▪ Bopton High - No
 ▪ Rivdale High - Yes  ▪ Rivdale High - No
 ▪ Rivdale High - No comment
 ▪ Bopton High - No comment

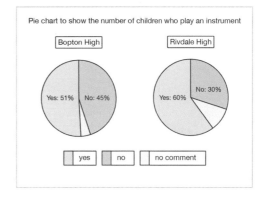

Pie chart to show the number of children who play an instrument

Bopton High        Rivdale High

Yes: 51%   No: 45%      Yes: 60%    No: 30%

yes      no       no comment

**Level 3:** *cont.*

**23.** The pie charts show the number of boys and girls
who attend two different schools. Which school
has more boys? Explain your answer.

a
b
c

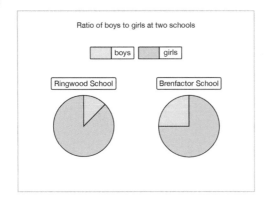

**24.** A golf club has 243 members and represents the
age of these on a pie chart. Explain what mistake
has been made.

a
b
c

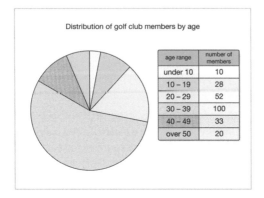

**25.** A survey asked a number of people to name their
favourite holiday destination. If 17 people said
Mexico, how many people were surveyed in total?

1
2
3

 68

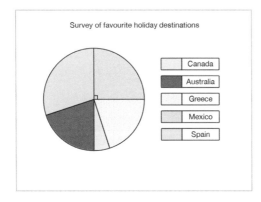

---

**Level 4:** Problem Solving - Comparing data between
pie charts and calculating angles.

✿ **Required:** 5/5   ✿ **Student Navigation:** on

✿ **Randomised:** off

**26.** An ice-cream van records how many ice-creams
are sold in a day. In degrees, what **angle** should be
used for the section for vanilla ice-creams?
*Don't include the units in your answer.*

1
2
3

▪ 180

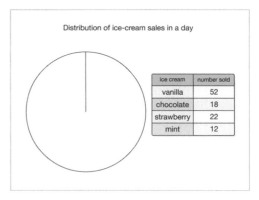

**27.** Use the data shown on the pie charts to work out
how many more children have a brother in class 7f
than in class 7d.

1
2
3

▪ 5

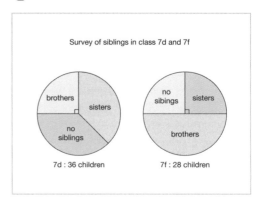

**28.** The results of a survey are recorded. In degrees,
what **angle** should be used for the section for The
Y Factor?
*Don't include the units in your answer.*

1
2
3

▪ 32.4

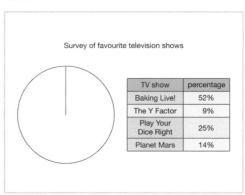

---

**29.** The pupils at Hex Academy are asked which planet they would most like to visit. 33 choose Mars. How many pupils are surveyed altogether?

1
2
3

▪ 220

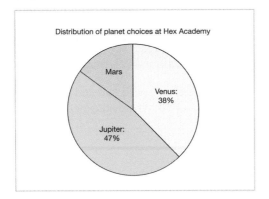

**30.** At a national sports competition, how many more boys competed in the **hurdles** than girls?

1
2
3

▪ 65

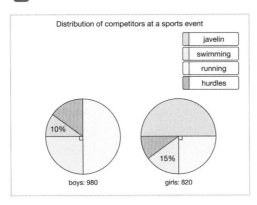

# Calculate the mean

**Competency:**    Describe, interpret and compare observed distributions of a single variable through: appropriate graphical representation and appropriate measures of central tendency (mean, mode, median) and spread.

**Quick Search Ref:**    10332

**Correct:** Correct.    **Wrong:** Incorrect, try again.    **Open:** Thank you.

**Level 1:**    Understanding - Calculating the mean from a set of data.

✸ **Required:** 6/8    ✸ **Student Navigation:** on    ✸ **Randomised:** off

---

**1.** The mean is found by:

1/4

- sorting all the numbers in a data set into ascending order and identifying the middle number.
- identifying the most common value in a data set.
- adding together all the numbers in a data set.
- adding together all the numbers in a data set and dividing the sum by how many numbers there are.

**2.** What is the mean value of the following data set?

3, 4, 4, 5, 6, 7, 8, 8, 8, 9.

▪ 6.2

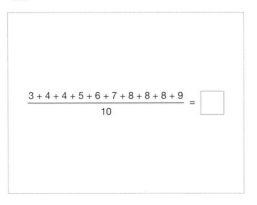

$$\frac{3 + 4 + 4 + 5 + 6 + 7 + 8 + 8 + 8 + 9}{10} = \boxed{\phantom{00}}$$

**3.** What is the mean age of the youth club members?

▪ 13

| name | age |
|------|-----|
| Jane | 14 |
| Sam | 12 |
| Adil | 12 |
| Jack | 13 |
| Helen | 13 |
| Jo | 14 |

**4.** Mr Farran's class have a quiz and each group gets a score out of 50. What is the mean result for the class?

▪ 30

| group 1 | group 2 | group 3 | group 4 | group 5 | group 6 |
|---------|---------|---------|---------|---------|---------|
| 27 | 38 | 19 | 42 | 32 | 22 |

**5.** The athletics team record the number of laps they run in 30 minutes. What is the **average** number of laps for the team?

▪ 31

| athlete | laps in 30 mins |
|---------|-----------------|
| Kieran | 33 |
| Elliot | 34 |
| Jessica | 28 |
| Laura | 31 |
| Stefan | 32 |
| Dan | 30 |
| Kay | 29 |

**6.** A survey records how long a group of 8 teenagers spend playing computer games in one day. Which **two** calculations will give the mean average for the group?

2/6

- ■ (90 + 210 + 180 + 180 + 90 + 60) ÷ 8
- ■ (90 + 210 + 30 + 180 + 180 +120 + 90 + 60) ÷ 8
- ■ (90 + 210 + 30 + 180 + 180 +120 + 90 + 60) ÷ 6
- ■ (90 + 210 + 30 + 180 + 180 +120 + 90 + 60) × 8
- ■ (90 + 210 + 90 + 60) ÷ 4  ■  960 ÷ 8

| name | time gaming (mins) |
|---|---|
| Louise | 90 |
| Ben | 210 |
| Andy | 30 |
| Callum | 180 |
| Jack | 180 |
| Sian | 120 |
| Anna | 90 |
| Mike | 60 |

**7.** Simon runs every day for 10 days. What is the average distance of his training runs?
*Include the units km (kilometres) in your answer.*

- ■ 7.4 km  ■  7.4 kilometres

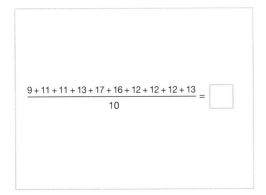

|  | runs (km) |  |  | runs (km) |
|---|---|---|---|---|
| Monday | 4 | ○ ○ | Monday | 4 |
| Tuesday | 7 |  | Tuesday | 7 |
| Wednesday | 7 |  | Wednesday | 9 |
| Thursday | 6 |  | Thursday |  |
| Friday | 3 |  | Friday |  |
| Saturday | 12 |  | Saturday |  |
| Sunday | 15 | ○ ○ | Sunday |  |

**8.** What is the mean value of the following data set?
9, 11, 11, 13, 17, 16, 12, 12, 12, 13.

- ■ 12.6

$$\frac{9 + 11 + 11 + 13 + 17 + 16 + 12 + 12 + 12 + 13}{10} = \boxed{\phantom{x}}$$

**Level 2:** Fluency - Finding and using the mean in context (including charts).

✱ **Required:** 6/8    ✱ **Student Navigation:** on

✱ **Randomised:** off

**9.** A box of assorted biscuits is labelled 360 g. If there are 18 biscuits in the box, which calculation will give the mean mass of one biscuit?

1/4

- ■ 360 g × 18  ■  360 g ÷ 18  ■  18 ÷ 360 g  ■  360 g - 18

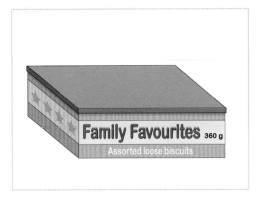

**10.** Bev's hens lay 45 eggs in three months. Calculate the mean number of eggs laid each month.

- ■ 15

**11.** A shop has three bags of apples left at the end of the day. From these remaining bags, calculate the mean weight of an apple.
*Include the unit g (grams) in your answer.*

- ■ 25 g  ■  25 grams

| number of apples | total weight (g) |
|---|---|
| 4 | 115 |
| 6 | 145 |
| 10 | 240 |

**12.** Jasmin buys a new book each week in the summer holidays. What is the mean cost of the books she buys?

**a b c**

*Include the £ sign in your answer.*

■ £7    ■ £7.00

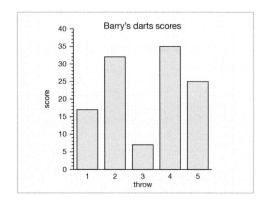

| week | cost of book (£) |
|------|------------------|
| 1 | 8.00 |
| 2 | 7.00 |
| 3 | 5.00 |
| 4 | 9.00 |
| 5 | 6.00 |
| 6 | 7.00 |

**13.** Barry is playing darts and puts his first five scores into a bar chart. What is the mean of his scores?

**1 2 3**

■ 23.2

*Barry's darts scores (bar chart)*

**14.** In a class of 30, the children earned a mean average of 6 gold stars during one week. **How many stars in total** were awarded that week?

**1 2 3**

■ 180

**15.** Ed earns £520 over one year. What is the average amount he earns each week?

**a b c**

*Include the £ sign in your answer.*

■ £10    ■ £10.00

**16.** Geography students recorded a total rainfall of 270 millimetres (mm) during April. What was the mean daily rainfall?

**a b c**

*Include the units mm (millimetres) in your answer.*

■ 9 millimetres    ■ 9 mm

Level 3:    Reasoning - Missing numbers, inverse operations, interpretation of mean.

✴ Required: 5/5    ✴ Student Navigation: on
✴ Randomised: off

**17.** The mean average mass of a carrot is 50 grams (g). If they are sold in 500 g bags, how many carrots are in each bag?

**1 2 3**

■ 10

**18.** The mean score in a set of 10 maths tests was 56. What is the missing score?

**1 2 3**

■ 58

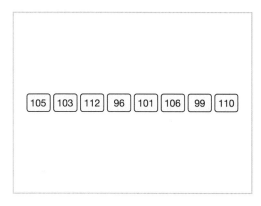

**19.** Grace says, "To find the mean of the following numbers, I can round each number to 100, add them together and divide by 8".

**a b c**

Will Grace's answer be accurate? Explain your answer.

105, 103, 112, 96, 101, 106, 99, 110.

| 105 | 103 | 112 | 96 | 101 | 106 | 99 | 110 |

**20.** Which **five** prices have a mean of £4.50?

☐
☒
☐

■ £2.40  ■ £9.15  ■ £5.10  ■ £7.25  ■ £1.20  ■ £3.50

■ £4.25

5/7

**Level 3:** *cont.*

**21.** Claire says she has done better than average in Maths, but worse than average in English. Is she correct? Explain your answer.

a
b
c

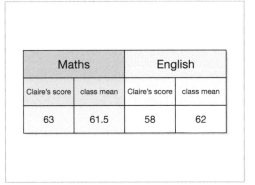

| Maths | | English | |
|---|---|---|---|
| Claire's score | class mean | Claire's score | class mean |
| 63 | 61.5 | 58 | 62 |

**Level 4:** Problem Solving - Multi-step calculations in context.

✱ **Required:** 5/5    ✱ **Student Navigation:** on
✱ **Randomised:** off

**22.** Which planet's diameter is closest to the mean diameter of all the planets?

☐
☒
☐

1/7

▪ Mercury ▪ Venus ▪ Mars ▪ Jupiter ▪ Saturn
▪ Uranus ▪ Neptune

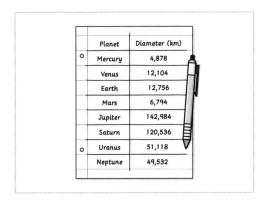

| Planet | Diameter (km) |
|---|---|
| Mercury | 4,878 |
| Venus | 12,104 |
| Earth | 12,756 |
| Mars | 6,794 |
| Jupiter | 142,984 |
| Saturn | 120,536 |
| Uranus | 51,118 |
| Neptune | 49,532 |

**23.** Amy lives with her Mum, Dad and three brothers. Each member of the family has a job and their mean earnings are £12,000 per year. When Amy moves out, the mean earnings of the household increase to £12,400. How much does Amy earn?
*Include the £ sign in your answer.*

a
b
c

▪ £10,000 ▪ £10000

**24.** What is the mean of the five card values?

1
2
3

▪ 4.56

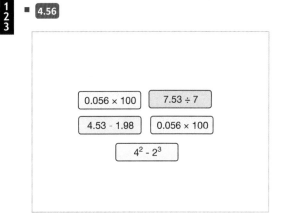

| 0.056 × 100 | 7.53 ÷ 7 |
|---|---|
| 4.53 - 1.98 | 0.056 × 100 |

$4^2 - 2^3$

**25.** John enters an 8 mile run. He runs 2 miles at 6 miles per hour (mph) and 1 mile at 4 miles per hour.
On average, how many minutes does John need to complete each remaining mile to finish in 75 minutes?
*Don't include units in your answer.*

1
2
3

▪ 8

**26.** In a sunflower growing competition, the team with the highest mean average height wins. By how many centimetres (cm) does Helen's team win?
*Include the units (cm) in your answer.*

a
b
c

▪ 18 centimetres ▪ 18 cm

Alfie's team                Helen's team

1.2 m  131 cm  970 mm        180 cm  103 cm  1.3 m  1,230 mm

# Calculate the mean, median, mode, range and outlier

**Competency:** Describe, interpret and compare observed distributions of a single variable through: appropriate graphical representation and appropriate measures of central tendency (mean, mode, median) and spread.

**Quick Search Ref:** 10150

Correct: Correct.      Wrong: Incorrect, try again.      Open: Thank you.

**Level 1:** Understanding - Calculating the mean, median, mode and range of a data set.

✿ **Required:** 7/10            ✿ **Student Navigation:** on            ✿ **Randomised:** off

---

1. What is the **mean** value of the following data set?
   3, 4, 4, 5, 6, 7, 8, 8, 8, 9.

   ▪ 6.2

   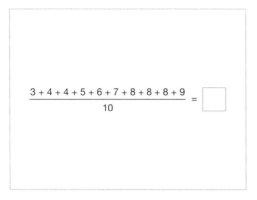

   $$\frac{3 + 4 + 4 + 5 + 6 + 7 + 8 + 8 + 8 + 9}{10} = \boxed{\phantom{x}}$$

2. Find the **mode** of the following data set:
   3, 4, 4, 5, 6, 7, 8, 8, 8, 9.

   ▪ 8

   3 4 4 5 6 7 8 8 8 9

3. Calculate the **range** of the following data set:
   3, 4, 4, 5, 6, 7, 8, 8, 8, 9.

   ▪ 6

   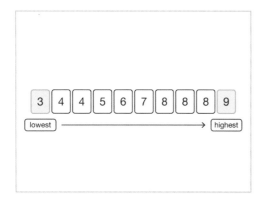

   3 4 4 5 6 7 8 8 8 9

   lowest ────────────→ highest

4. What is the **median** value of the following data set?
   3, 3, 4, 4, 6, 7, 7, 8, 8, 8, 8.

   ▪ 7

   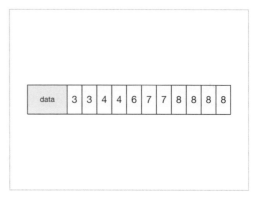

   | data | 3 | 3 | 4 | 4 | 6 | 7 | 7 | 8 | 8 | 8 | 8 |
   |------|---|---|---|---|---|---|---|---|---|---|---|

5. Eight children in class 6 record their height in a table. Select the two values for the **mode** height.

   ▪ 127  ▪ 132  ▪ 133  ▪ 134  ▪ 135  ▪ 137

   2/6

   | name | height (cm) |
   |------|-------------|
   | Danny | 137 |
   | Luke | 132 |
   | Nathan | 127 |
   | Louise | 133 |
   | Helen | 135 |
   | Isaac | 137 |
   | Maya | 133 |
   | Joseph | 134 |

6. Which value is the **outlier** in the following data set?
   6, 86, 92, 94, 96, 96.

   ▪ 6

   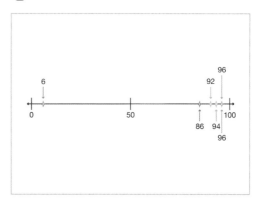

**7.** There are 10 players in a netball squad. The coach records the number of games each girl plays in the season. What is the **median** number of games played by each girl?

1/4    18, 18, 14, 15, 14, 17, 13, 15, 18, 17.

   ▪ **15**  ▪ **16**  ▪ **17**  ▪ **18**

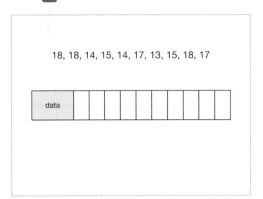

**8.** Mr Farran's class have a quiz and each group gets a score out of 50. What is the **range** of the results?

   ▪ **23**

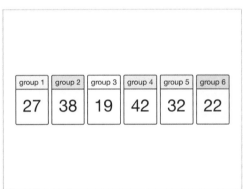

**9.** Find the **mode** of the following data set:
12, 8, 9, 12, 15, 13, 15, 19, 15, 5.

   ▪ **15**

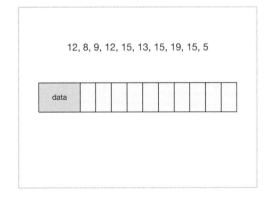

**10.** What is the **median** of the following data set?
15, 8, 9, 12, 13, 15, 16, 15, 5.

   ▪ **13**

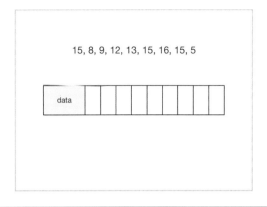

**Level 2:**  Fluency - Finding the mean, median, mode and range in context (including charts).

✱ **Required:** 7/10    ✱ **Student Navigation:** on
✱ **Randomised:** off

**11.** Calculate the mean, median and mode of the following data set and select the one with the **highest** value:

1/3    8, 15, 6, 8, 9, 9, 8, 12, 10, 11.

   ▪ **mean**  ▪ **median**  ▪ **mode**

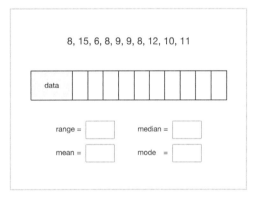

**12.** A shop has three bags of apples left at the end of the day. From these remaining bags, calculate the **mean** weight of an apple.
*Include the unit g (grams) in your answer.*

   ▪ **25 grams**  ▪ **25 g**

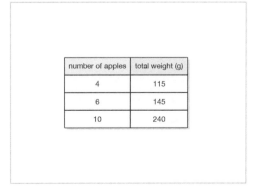

**13.** The bar graph shows the shoe size of each person who goes to a bowling alley for a birthday party. What is the **mode** shoe size?

- 6

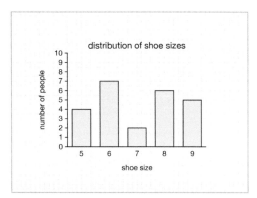

**14.** Gavin asked his friends how many books they bought last year. He can't remember one of their answers, but he knows the median is 7.5 books. What is the missing number?

- 8

**15.** Barry is playing darts and puts his first five scores into a bar chart. What is the **range** of his scores?

- 28

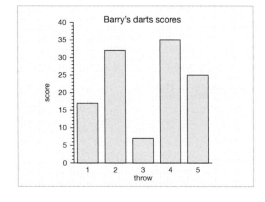

**16.** Franky wants to know what flavour of ice-cream is the most popular. What is the **mode** flavour of ice-cream?

1/5   ▪ strawberry ▪ vanilla ▪ chocolate ▪ mint ▪ raisin

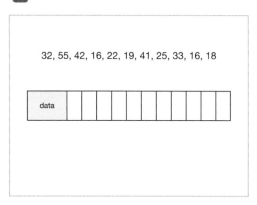

| flavour | ice-creams sold |
|---|---|
| strawberry | 13 |
| vanilla | 8 |
| chocolate | 17 |
| mint | 14 |
| raisin | 8 |

**17.** The chart shows the ages of the people who go to the cinema. What is the age **range**?

- 39

32, 55, 42, 16, 22, 19, 41, 25, 33, 16, 18

| data | | | | | | | | | | |
|---|---|---|---|---|---|---|---|---|---|---|

**18.** A doctor's surgery records the age of its patients. What is the **median** age?

- 45

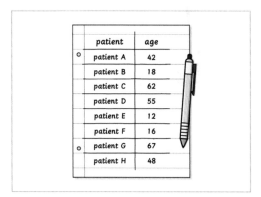

| patient | age |
|---|---|
| patient A | 42 |
| patient B | 18 |
| patient C | 62 |
| patient D | 55 |
| patient E | 12 |
| patient F | 16 |
| patient G | 67 |
| patient H | 48 |

**19.** Jasmin buys a new book each week in the summer
holidays. What is the **mean** cost of the books she
buys?
*Include the £ sign in your answer.*

■ £7.00  ■ £7

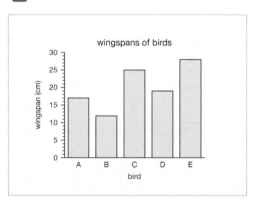

**20.** Wesley measures the wingspan of all
the birds that land in his garden in an afternoon.
What is the **range** of the wingspans?
*Do not include the units in your answer.*

■ 16

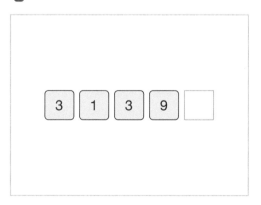

Level 3:  Reasoning - Negative numbers, missing
numbers and outliers.

✱ **Required:** 5/5   ✱ **Student Navigation:** on
✱ **Randomised:** off

**21.** A group of five numbers have a mean of 4, a
median of 3 and a mode of 3. What number is
missing from the group?

■ 4

**22.** The **range** of the following data set is 44 and the
missing number has 3 digits. What is the missing
number?
72, 72, 66, 82, 68, ?

■ 110

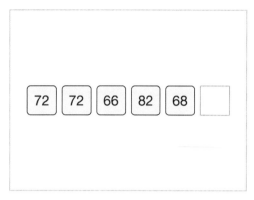

**23.** Alex's teacher calculates the mean score for a
maths test. What can Alex's teacher do to
calculate a more accurate average of the scores?

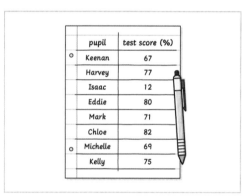

**24.** Grace says, "To find the mean of the following
numbers, I can round each number to 100, add
them together and divide by 8".
Will Grace's answer be accurate? Explain
your answer.
105, 103, 112, 96, 101, 106, 99, 110.

**25.** Danny and Eddy are supposed to have six number cards each but Danny has taken one too many. Which card can Danny give to Eddy so that they both have the same **range** of numbers?

1
2
3

▪ 7

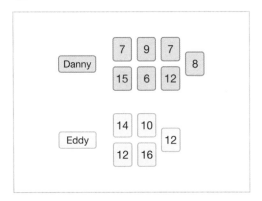

**Level 4:** Problem Solving - Calculating missing values using mean, median, mode and range.

✱ Required: 5/5    ✱ Student Navigation: on
✱ Randomised: off

**26.** Katie and her four friends each think of a number between 1 and 9. The sequence of the five numbers has the following rules:
The mean, median, mode and range = 5;
Only one appears twice;
The lowest number chosen is 2.
*Write the five numbers chosen by Katie and her friends as a number sequence, in the format, 1, 2, 3, 4, 5.*

a
b
c

▪ 2, 5, 5, 6, 7. ▪ 2, 5, 5, 6, 7 ▪ 2, 5, 5, 6, 7,

**27.** There are 46 football matches in one week. The number of goals scored in each match is recorded in a line graph. What is the **median** number of goals scored in a single match?

1
2
3

▪ 2.5

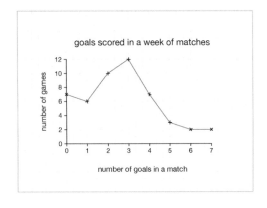

**28.** Amy lives with her Mum, Dad and three brothers. Each member of the family has a job and their mean earnings are £12,000 per year. When Amy moves out, the mean earnings of the household increase to £12,400. How much does Amy earn?
*Include the £ sign in your answer.*

a
b
c

▪ £10,000 ▪ £10000

**29.** The value of Y on the number line is 21. What is the **range** of X, Y and Z?

1
2
3

▪ 77

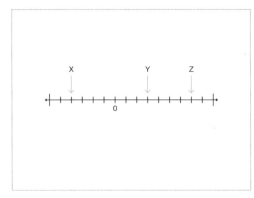

**30.** John enters an 8 mile run. He runs 2 miles at 6 miles per hour (mph) and 1 mile at 4 miles per hour.
On average, how many minutes does John need to complete each remaining mile to finish in 75 minutes?
*Don't include units in your answer.*

a
b
c

▪ 8

# LbQ Super Deal
## Class set of tablets and charging cabinet

Class charging
& storage cabinet

**+** **32 x**
Pupil 7" HD tablets
with protective
cover

**+** **1 x**
Teacher 10" HD tablet
with protective
cover

## *Special Offer Price

£1,100 per year on 3 years compliant operating lease

Subject to a £150 initial documentation fee

LbQ Question Set subscription required to be eligible

Min 1 LbQ subscription per set £200/year or £500/3 years

Learning by Questions App pre-loaded

Includes 3 years advanced replacement warranty on tablets (damage not covered)

Prices exclude VAT and delivery

Option to renew equipment or purchase at end of agreement

Available in United Kingdom and Republic of Ireland only

**Product Code TC001**

**£1,100***
per year for
3 years
including
warranty

# Specifications

## Charging & Storage Cabinet

- 32-bay up to 10" tablet charging cabinet
- 2 easy access sliding shelves – 16 bays on each shelf
- 4 efficient quiet fans for ventilation
- Locking Doors (4 keys)
- 4 castors / 2 x handling bars Overload, leakage and lightning surge protection
- CE / ROHS / FCC compliancy
- Warranty 3 years

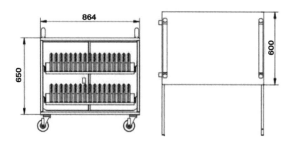

## Student and Teacher tablets configuration

All tablets with LbQ Tablet Tasks App pre-loaded and installed in cabinet including charging cables and mains adapters for quick and easy deployment when onsite in classroom.

| | 7" Android Tablet with Protective Cover | 10" Android Tablet with Protective Cover |
|---|---|---|
| **Display** | | |
| 1920 * 1200 IPS | ✓ | ✓ |
| 16:10 Display ratio | ✓ | ✓ |
| Capacitive 5-touch | Capacitive 5-touch | Capacitive 10-touch |
| **System** | | |
| Cortex 64bit Quad Core 1.5GHz CPU | ✓ | ✓ |
| 2GB of RAM | ✓ | ✓ |
| 16Gb of storage | ✓ | ✓ |
| Android 7.0 | ✓ | ✓ |
| Front 2M and rear 5M Camera | ✓ | ✓ |
| **Input / Output Ports** | | |
| 1 x Micro SD Slot | ✓ | ✓ |
| 1 x Micro USB (PC / device / charger) | ✓ | ✓ |
| Micro-HDMI output | ✓ | ✓ |
| 1 x Earphone, 1 x Speaker, 1 x Mic | 1 x Earphone, 1 x Speaker, 1 x Mic | 1 x Earphone, 2 x Speaker, 1 x Mic |
| **Communication** | | |
| Wifi – 802.11a/b/g/n | ✓ | ✓ |
| GPS module | ✓ | ✓ |
| Bluetooth | ✓ | ✓ |
| **Power** | | |
| 5V 2A | ✓ | ✓ |
| Battery | 3000mAh battery | 7000mAh battery |
| **Physical** | | |
| Colour: Metal black | ✓ | ✓ |
| Weight: | 230g | 560g |
| Dimensions: | 192 x 112 x 9mm (approx.) | 263 x 164 x 9mm (approx.) |
| **Warranty** | | |
| 3 years Advanced Replacement for faulty tablets - does not cover damage CE / ROHS / FCC compliant | ✓ | ✓ |